CHRONICLES OF THE GODS

BOOK TWO

LINDSAY SCOTT CISNEY

CITI OF BOOKS

CITIOFBOOKS, INC.
3736 Eubank NE Suite A1
Albuquerque, NM 87111-3579
www.citiofbooks.com

Hotline: 1 (877) 389-2759
Fax: 1 (505) 930-7244

Ordering Information:
Quantity sales. Special discounts are available on quantity purchases by corporations, associations, and others. For details, contact the publisher at the address above.

Printed in the United States of America.

ISBN-13:	Softcover	979-8-90124-286-5
	eBook	979-8-90124-287-2

Library of Congress Control Number: 2026921366

CONTENTS

Part 1: Empires

Part 2: Alternate Earths

Part 3: Mars Rescue

Part 4: Earth Rescue

Part 5: Building a Better World

Part 6: In the Lion's Den

Part 7: Becoming the Goddess of Terra Nova

It will be the greatest leap forward in human history, when somehow, we will be able to pass on and have it sink into human consciousness, both individually and collectively, the lessons of human behavior that our forebearers learned.

—L. Scott Cisney, October 2024

PART ONE

Empires

CHAPTER 1

Hatcher's Anguish

Benston Colony, September 2, 2354

"I WAS PERSONALLY SURPRISED at how affable the Khan was," Teddy Rodriguez said to Gunner.

"Yeah, he is a well-spoken man, actually quite handsome," Shelby Gee added.

Both Teddy and Shelby were among the armed guards who stood by Peter and Terra as they traded people.

"His people are vicious," Hatcher added, shaking his head.

"Perhaps his position as the leader of his people forces him to act in aggressive ways. Maybe he, personally, is not vicious. I hope Terra can moderate the behavior of his people," Teddy added.

"Well, it sounds like we will see them again once they sail their ships around the peninsula," Gunner mused.

The trip back to Benston was bittersweet for Hatcher, Elias, and Joshua. Hatcher had lost Terra to another man, Elias had lost his friend Jed, and Joshua had lost his younger brother, Thomas. Each sat in silence, with somber thoughts. After a couple of hours, they were back in Benston.

As Hatcher walked down the ramp, family members were there to greet the returning colonists.

Taylor walked up to Hatcher with Zach, Willie, and Connie. Each of the children were crying as they hugged Hatcher. Lastly, Taylor grabbed Hatcher's hand and pulled him close to give him a hug and a kiss.

As she did, she whispered, "We're going to get her back. I don't know how, but we will."

Hatcher could not help himself at this point as he began sobbing almost uncontrollably. Taylor hugged him tighter as he did this, trying her best to soothe him.

Gunner walked up to President Davis and motioned for him to separate from the gaggle of people so that they could talk more privately.

Gunner cleared his throat and said, "You should know that our conversation with the Khan went well. Whatever his people may lack in a civilized way, he seems to have it. He is returning with Terra and Peter on his ships. Terra invited him to overwinter down near the harbor. Peter is helping them navigate to the harbor. I'm not sure how long it might be before we see them again, probably a few weeks. We should discuss how we will help them with shelter and food."

"We've had a very good harvest this summer. Our provisions from the *Calypso* can be shared with the Khan's men, though with as many men as he has, those provisions will be mostly exhausted by spring.

We will need to determine, right away, how much in the way of supplies they already have. We are facing a big challenge in terms of feeding our people and theirs. We need to have Congress appoint a committee to oversee this situation. For now, I'd like you to draw up a list of people who should sit on this committee and have it ready for me to give to the congress when they convene on Wednesday. Can you do that for me, Gunner?"

"Yes, of course, Mr. President," Gunner answered.

CHAPTER 2

Camp of the Horsemen

THE JOURNEY BACK to the horsemen's camp ended up taking two days. The trail had become covered with snow from an early-season storm that erased the trail in the hill country that they passed through. They overnighted at the top of the pass. It was night when they stopped and put up tents for the night. The Khan invited Terra to share his bed. "It will be cold tonight. Come lie with me. I will keep you warm." Terra accepted and removed her clothes and slid into bed next to the Khan, who was also naked. Terra began crying and shaking slightly as he put his arm around her, and they spooned to stay warm.

"We do not need to make love tonight, Terra. I know this is hard for you. I will do everything I can to make you happy, my sweetheart."

Terra rolled over to face him, tears still running down her cheeks. "I know you have a good heart, my Khan. I can love you for that now. I will work hard to support you in all that you do. I wish only that you would take counsel from me from time to time. I believe your people can become more peaceful and that by doing so, your name will be greatly honored, my Khan."

"Please, when we are alone, call me Ammon. You are my wife and will be the mother of our children. I am your husband and will be the father of our children. We are a family now."

Terra moved closer to the Khan and kissed him gently on the lips. As she did so, she ran her hand down the muscles on his chest and stopped just short of his manhood. She smiled and laughed softly as the Khan let out a soft moan of anticipation. He leaned closer and picked her up and placed her on his abdomen. He began softly caressing her breasts. As he did so, Terra began moaning with pleasure. Terra held his face in between her hands and kissed him again, this time deeply. The Khan rolled Terra onto her back and began making love to Terra as she cried out in pleasure. When they were finished, Terra lay down next to him, and they both fell asleep.

Morning came, and one of the horsemen came to the door of the tent to wake up his Khan and his new bride. Both Terra and Ammon had slept deeply after making love. The grogginess they both felt left them as they got out from under the covers, and the cold morning air hit them.

As Terra dressed, he watched her and smiled. "You are beautiful, in addition to being an oracle. I am the luckiest man alive."

Terra blushed and turned towards him. "Ammon, you are too kind, though I wonder if your eyesight is failing you!"

She laughed as she said this and walked over to him, still nude, and kissed him. For a moment, they both considered, silently, if they had enough time.

Terra sighed and said, "We really must get going, Ammon."

"Yes, I suppose we must," he said.

They descended from the hills by mid-morning and were on a wide plain that led to the sea, which could be seen far in the distance. They traveled another hour, crested a final hill, and stopped. Before them, a large encampment lay below. As they came near the edge of the camp, a group of horsemen rode up and greeted their returning leader.

"Great Khan, welcome back. I trust your mission was successful?" an older, bearded man on horseback asked.

"Yes, Allan. You see, here is my new wife, the oracle named Terra," the Khan said, pointing to Terra, who was on horseback on his right side.

"What are your orders, Great Khan?" Allan asked.

"We will break camp the day after tomorrow and head back to the ships. I have found a place for us to winter next to our new friends, the people of Terra who came from afar across the night sky," the Khan said with a wave of his hand upward toward the darkening sky. "Make ready my tent and a meal for myself and those with me. We have ridden hard today to be here." The riders nodded their heads and rode back toward the camp.

As they entered the camp, the Khan stopped to talk to a great many people. It turned out that there were women and children in the camp, though not nearly enough to represent the families of every man present.

"Where is home?" Terra asked.

"Originally, we came from the high grasslands that are a great distance from here, —north and east," Ammon answered.

"Will we ultimately be returning there?" she asked.

"That is my decision, and I have not yet decided. We came from a land somewhat poor in resources. The only way my peoples could survive and thrive was by plundering the wealth and peoples that we encounter, and we are very good at it," he answered, now clearly embarrassed as he saw the look of empathy and sadness in Terra's eyes. "I hope we can settle down soon, learn to live off a better land," he finished.

"Do you know what farming is, Ammon?" she asked.

"Yes, we have seen small groups attempting to grow food from the ground. It never seems to provide them a better life than what we have," he answered.

"I can help you and your people with farming. Farming is the way to feed large sedentary groups of peoples and build great civilizations that bring order, prosperity, and justice to the very wild and sometimes unforgiving world around us. I want to help you become the great leader of this new civilization, Ammon," Terra said as she smiled and reached over to grab Ammon's hand.

Ammon smiled back at her and just stared at her for a moment, and then said, "I need to remember every day how lucky I am that you are here. Our tent awaits, my love!"

The tent was better described as almost a house. Entrance to the tent was through a doorway made of timbers that were part of an elaborate wooden structure that securely held up the animal-skin walls and roof. The floor was comprised of wooden planks covered in many locations with elaborate and beautiful rugs. An animal-skin bladder full of water provided drinking and cleaning water for the tent occupants. A large tub of steaming hot water was also in the tent, along with two younger females, probably sixteen years old, who were there to assist the person in the tub. Ammon told Terra that the two younger females were to be Terra's personal servants.

"I have no need of servants, Ammon, though I appreciate the kindness that begat the idea," Terra said, now clearly embarrassed at the trappings and perks of being wife to the Great Khan.

"Terra, you may do as you wish, but life here will be much harder than you have experienced. If you change your mind, the girls will always be available to help you. This was a great honor for their fathers to see them serve you. They may come to believe that their daughters displeased you or me in some way. Sometimes the treatment of daughters by their fathers is harsh. Perhaps you should consider this," he finished.

Terra was silent for a moment and then said, "Have them stay. I'm sure I can find them of use."

Terra was reaching a firmer idea of how large the job in front of her would be. The treatment of women that became the norm on Earth for thousands of years would happen here if she failed. That failure would doom this world to the centuries of bloodshed and strife that had beset Earth. Terra did not intend to fail.

CHAPTER 3

Roller Coaster Ride

THE SHIPS WERE much smaller than she imagined. The horses actually had the best accommodations, Terra observed. They were put below the deck in stalls and fed and watered well. At any time, the deck of the ship had two or three horses being led around by the horse handlers. The people slept on deck, sometimes under large animal skins that sprang up when it began to rain. For her "quarters," Terra had the convenience of a more permanent structure—a tent erected on the deck near the rear.

It struck Terra that everyone seemed happier to be back on the sea. She had noticed the same when they were on horseback, on the trail back to their old encampment. To Terra, everything she was experiencing seemed as if it were a hardship, yet many of these people were making do and were happy.

On the seventh day out and with another ten days in front of them, the weather turned bad. The wind started blowing from the south, which meant all the ships were sailing against the wind. The ocean began to experience large rolling swells. Terra and Peter were both seasick. The two of them spent most of their time in Terra's quarters. Peter had queried the weather cubes that the *Calypso* had put into orbit before it left to give an indication of approaching weather systems for the colonists. The data showed a circulating storm approaching them from the south. Peter wasn't sure if this counted as a hurricane, but he supposed it did.

* * *

"Yes, I'm saying there is what appears to be a protected harbor that we can reach by evening, today. We should go there and wait this storm system out," Peter said, exasperated that the Khan still thought it would be safer out on the open sea versus anchored in a safe harbor.

"The sea will push into the harbor. If we anchored there, we would become marooned there. The ships would be thrown up onto the land," the Khan said calmly.

"Then perhaps we should use the strong headwind we are sailing against as a tailwind and sail back the direction we came. The storm could turn east or west and possibly leave us out of its track," Peter countered.

"We will proceed on our southerly course," the Khan said sternly.

This conversation had been at midmorning. By noon, the sun was very dim due to bands of rain. At about 1:00 p.m., a strange ship was seen southeast of the Khan's boat. The Khan signaled two of the closest of his ships that he wanted them to investigate this strange ship. An hour later, one of his ships pulled alongside the

Khan's ship and brought a man aboard. The man spoke a strange dialect, but one of the women on board—a woman captured on an earlier raid—spoke that dialect and was brought to the man to translate what he was trying to say.

"He say big storm destroyed his friend's ships. The boat he on OK, but the mast was broken, and they want to come aboard and serve the Khan as they in bad way," the woman translated in her broken version of the horsemen's language.

Terra, who had been listening, said, "We must sail west, my Khan. The storm will turn east and make landfall. We must do it now, or many of us will perish. I say this as an oracle."

The Khan, Peter, and the other men had gathered to discuss the situation. They all stood silent, staring at Terra, then switched their gaze to the Khan to see what he would say.

"We will sail west, as Terra says," the Khan responded.

Sailing west took them away from the continent. The Khan and his people were nervous about this, as no one had ventured away from the sight of the west coast of the continent. What lay across these great waters was a mystery for the Khan's people. They sailed for a full day and a half westward. At that point, Terra said it was safe to return to the vicinity of the coast they had left behind.

With the coast in sight, they continued to hug the coast around what Peter was calling the Iberian peninsula. The naming of the peninsula and a great many features started to take on Earth names as the colonist wanted familiar names for this decidedly alien world. On the twelfth day, they rounded the southern tip of the peninsula and began sailing north, with favorable winds, toward the Benston Colony. On the seventeenth day, the massive armada of ships began arriving in the Benston Harbor. The harbor proved to be inadequate for the number of ships, as many ships were anchored in coves up and down the coast from Benston.

CHAPTER 4

Supply Dilemmas

"IN SUMMARY, WE have eleven months of food for a population of 27,000. This is based on an average caloric intake of 2,500 calories for an adult male with sixty-three grams of protein per day. Since approximately 77% of the Riders are adult males, this represents a conservative estimate. Women and children consume fewer calories and less protein," said Dr. Bogdan Dozic, a logistics expert.

"Since we are at the end of the main growing season, and harvesting new crops is ten or eleven months away, this means we would be cutting it close. The Riders are excellent hunters and have some food provisions already. We can assume this also makes our estimate even more conservative," Terra added.

"Yes, that is correct," Bogdan answered.

"What are your winters like here?" the Khan asked.

"The winters rarely result in freezing conditions. It does rain on many days, however," Carrie said.

"We are building housing units for your people, but it may take all winter to build enough for all of your people," President Davis noted.

The Khan shifted in his seat and then said, "Yes, thank you, that will be much nicer than we are accustomed to. Our tents will be adequate for most of my people, though perhaps the women and children should be put in your housing first. We have an unusual amount of sickness at present. One of your healers described the illness as pneumonia or bronchitis. They are having those affected take medicines—I am not sure what they called it."

"Targeted immunotherapy enhancers, or TIE. Great Khan, we can expect an increase in illnesses among both our peoples as neither group will have immunity to the diseases common to each group. In our history, many native populations that encountered new people suffered greatly from new viruses or illnesses. I am currently running scans on a sample of one hundred of your people to determine what illnesses your people are not accustomed to. So far, it looks like your people have deficiencies to most of the childhood viruses that we no longer experience among our people. We are producing immune boosters for your people, which should help prevent many of those diseases," Dr. Maria Senegal noted.

"A couple of years back, we experienced a terrible plague. Our healers had no medicines that helped. I hope we do not cause your people to become afflicted by this," the Khan said.

"That illness is almost identical to an Earth illness called the Bubonic Plague. We can treat that illness quite effectively. The illness is passed from rodents to people, so eliminating the rodent population will help greatly. In the meantime, we want to introduce sanitation measures among your settlement that should help reduce illness and disease, in general, greatly. We want you to understand why these measures work so we will begin educating your people on this topic," Dr. Senegal stated.

After the meeting, Terra told the Khan she was going to visit her children. When she got to her house she let herself in and found Taylor feeding the children. Terra's twins were a year and a half old now. Zach, Taylor's son, was three years old. All four were eating lunch. Taylor was trying to eat in between feeding scoops of food to Willie and Connie.

"Terra! How are you?" Taylor said with surprise and genuine happiness as she looked up and saw Terra enter the kitchen.

"Good! How are you? Let me help you with the kids."

"Sure thing. What are you doing here in Benston?" she asked.

"There was a meeting on food allotment and health issues. Ammon and I were attending. It's over now; we're riding back to the camp in a couple of hours. I wanted to see how you were doing. Is Hatcher here?"

"No, he went hunting with Joshua. The colony is trying to get a ton of deer and buffalo meat salted before the rainy season. Seems like everyone has a turn with the hunting parties, but Hatcher likes it, he has been out every day this week," Taylor said.

"How is he doing?" Terra asked.

"Honestly, I don't know. He's quiet now. He doesn't laugh. It seems like he is trying to keep himself busy to avoid thinking about all of 'this'. I wish he would talk to me about how he feels."

"I've seen him like that once before. There was a time when we almost broke up. He became quiet then, but eventually he came back. Of course, it helped that we got back together. I am so worried about him!"

Terra now began sobbing. Taylor set down the spoon that she had intended for Connie and came over to Terra and put her arm around her, and now she began crying as well.

"I think the kids are finished. Maybe you can help me put them down for their naps?" Taylor said in between sobs.

After putting the kids down for their naps, Terra and Taylor ended up back at the kitchen table.

"What's he like, the Khan?" Taylor asked.

"He's not at all like what I thought he would be like," Terra answered. "He's kind to me. I think his 'public' persona is 'tough' because that is what his people expect of him. He doesn't really like being that way. I hope, in time, that his people will change and become more like the way he really is—you know, kinder, more refined."

Taylor asked, "Will Ammon and his people be returning to their homelands or will they stay here?"

"I don't know what the long-term plan is. They came from a great distance to the north and east of here.

I believe their plan was to return there. Now, with a settlement going up, I'm hoping they will stay here, change their lifestyle, become farmers, and build a city-state. This is how things happened on Earth long ago. People started out hunting and foraging. Then they learned to grow crops and domesticate animals. You seem to have adapted well to this lifestyle!" Terra said with a smile.

"Well, it is much easier, and I don't mind that!" Taylor said with a chuckle.

The two hugged again, and now Terra kissed Taylor. The two stared at each other for a moment and then Taylor said, "I think we have time before the kids wake up!"

CHAPTER 5

What Has Become of the *Calypso*?

THE LOSS OF communication with the *Calypso* caught Sam by surprise. Everything seemed to be going well. The efficiency improvements on the ion propulsion system were going to shave four years off the voyage to Mars and four years on the way back.

There were no hints of problems. What went wrong? Sam thought.

As soon as the com loss occurred, they had sent out a light-speed message. They speculated that the *Calypso* EPC particle containment vessel or the EPC environmental control unit, designed to keep the particles super cooled, had failed, causing a loss in EPC capability. Of course, the conventional light- speed message was going to take years before they would get a reply. No one was holding their breath for a return message.

Sam wanted to talk to Terra, hoping she could ask Eve and find out what was going on.

Surely Eve will know what is going on, Sam thought.

He was going to head over to the Riders' encampment after his shift and find out what she knew. It had only been two days since com loss, and he wasn't sure if she knew about the situation.

Sam decided to walk down to the encampment, which was a good thirty-minute walk from Benston. The most direct route was not the most scenic, and Sam was not in a hurry, so he walked along the wide sandy beaches over to the encampment. This meant that the walk was now a forty-five-minute walk.

The early evening sun made everything seem clearer, more "alive." Sam liked this part of the day the most. The Terran moon—also a large natural moon, like Earth's moon—was just appearing in the sky owing to the slightly dimming light. The moon looked to be full, meaning his walk back along the shore would be even more beautiful than his walk to the encampment. At this time of day, the shortening days of autumn left a slight coolness to the air, which made him feel refreshed.

As he approached the encampment's lumber walls, two guards hailed him and came out to check him out. As they came to him, they spoke in a Terran tongue, which Sam did not understand. Not everyone spoke English yet, though there was a significant effort underway to get everyone proficient in the now galactic language.

Sam surmised their intent and said in a simple English phrase, "I [pointing to himself] am Terra's father. I have come to see her."

The two looked at each other for a moment, then brought him through the gate, and they hailed another Terran. A conversation ensued in their tongue, then it ended, and the two pointed at the third man, indicating Sam should go with him.

As they wound their way through the wooden huts of the encampment, Sam took note of the fact that everyone seemed to have a job to do, even children. He passed a group of women, who, along with several children, appeared to be mending fishing nets. Sam knew that the Riders were trying to catch large amounts of fish in the Terran Sea. The need for protein in their diets demanded that all manner of protein be harvested and salted away before the winter months, when some of the fish migrated south. They passed another group of men who were training young children how to fight with swords. Sam noted that both boys and girls were included in this training.

Well, that's a good sign. At least they are starting out by treating boys and girls the same, Sam thought.

Presently, they arrived at a very large wooden structure. Sam had been here only once before, and he recognized this as the Khan's "palace," though it was a modest one, at least at this time.

"Daddy! How are you?" Terra said as she came out of the large double doors at the front of the building.

"I'm doing well. How are you?" Sam answered back.

Sam noticed that Terra looked pregnant, perhaps three or four months along.

"At this time of day, I'm feeling pretty good. Earlier today, I wasn't feeling so great," she said with a wide smile.

"Well, you look good—radiant even, I'd say," Sam said back to her with a smile.

"As you may have guessed, I'm pregnant again. Dr Harrison says it will be a boy, the new Khan," Terra said, smiling again.

"Congratulations!" Sam said.

"So what is going on with the *Calypso*?" Terra asked.

"How did you know to ask?" Sam asked somewhat, but not totally surprised, since his daughter always seemed to know things at least as soon as everyone else, and usually before.

Sam answered her, "Two days ago, they failed to make the com cycle check in. Everything had seemed to be going well. The voyage was even going to be shortened by eight years due to improvements in the ion propulsion system. Then, as I said, two days ago, they just never checked in for their morning call. I was hoping you could tell me what happened, you know, maybe you or Eve know what has happened. At least I was hoping for that," he said.

"Well, let's go ask her," Terra said, sounding a bit concerned.

As they walked back to her private quarters, Terra said something in the Terran tongue to two women who were waiting for Terra at the door to her quarters. The two women stood up and left the large dining area, shutting the doors behind them as they did. Sam and Terra were now alone. Terra motioned for her father to follow her into her bedroom. They both sat down on the floor on a large animal hide rug, sitting cross-legged. Terra took a deep breath in and exhaled, then she began singing the Mother's Song. Sam listened and watched as his daughter's beautiful voice sang the words of the poignant song. Sam himself became overcome by the song, and soon he found himself singing. After just a moment of singing, the room seemed to become utterly and completely silent, then Eve appeared, standing on the rug between them.

"Mother Eve, I am so happy to see you!" Terra said with real excitement and love.

Eve smiled at them both.

"Mother, the *Calypso* is not returning coms. We are concerned for their well-being. Can you tell us what has happened to them?" Terra asked.

"They are alive and well, my daughter, though they have slipped backward 12,300 years along the time axis in the spacetime continuum. They will be at the ancient Earth within a few months, by the end of this year. They do not know how to re-enter the spacetime continuum where they left it and are seeking my assistance. This was all known ahead of time. The Ones needed the *Calypso* to reach the Earth's solar system sooner because the AOW knows of the rescue mission and were planning to destroy the *Calypso* and all remaining Earth people in the solar system. I will help them return to the appropriate time in spacetime so they can rescue the survivors on Mars and on the Earth," Eve said.

"There are survivors on Earth?" Terra asked.

"Yes, including both sets of your grandparents in Denver," Eve answered.

"I will get to see my grandparents again?" Terra asked excitedly.

"Yes, my precious daughter," Eve smiled as she said this.

For his part, Sam was also smiling and crying, hearing now that his and his wife's parents were alive and would be joining them on Terra Nova.

"Mother Eve, thank you for watching over our friends and family," Terra said with tears freely flowing down her cheeks.

Eve smiled at both of them, nodded, then she was gone.

CHAPTER 6

Dinner and a Proposal

WHEN THEY WERE finished, Terra and Sam went back into the dining room, where they found Tarish and Soom, the two young women who earlier had closed off the dining room and bedroom area for Terra and Sam when they talked to Eve.

Tarish, the shorter of the two, went over to Terra and whispered something into her ear. A muffled exchange followed between the two then Tarish left the room quickly.

"Daddy, the Khan has some special event this evening and—"

"That's ok. I need to get back home and help your mother with the kids. I'm glad you were able to talk to Eve. It helps to know that the *Calypso* is ok. Come by and see your mom and the kids sometime," Sam said as he excused himself, following Soom, who offered to escort him to the gate.

Terra sat on a chair in the dining room and began crying. Whether her tears were tears of sadness or happiness was not apparent to the Khan, Hatcher, and Taylor, who had walked in silently, unnoticed by Terra.

Taylor came over to Terra and put her hand on her shoulder, and asked softly, "Are you ok?"

"I'm fine. I'm so happy to see all of you, as she stood and wiped her tears and turned to face the three of them.

Terra decided to "listen in" to Ammon's thinking, who was standing next to Hatcher. Ammon had a sly smile on his face as he looked at Terra.

"Dinner, a surprise, a 'proposal'..." but that is all Terra got from Ammon's thoughts, who now spoke to her, "Terra, we have guests for dinner. We saw your father and Tarish on the way in. He seemed 'relieved.' Perhaps we can talk later about what he wanted. At any rate, I've invited your husband, Hatcher, and his second wife, Taylor, for dinner."

Terra froze at the word "husband". She looked at Ammon, who still had a smile on his face and there was something else. *Was it something funny?* she wondered.

If she could only disengage from a verbal conversation with everyone, then she could pick through Ammon's thoughts some more, but there was no break in the conversation.

Several servants started setting the table and arranging the chairs. This was followed by others who brought in a dinner fit for a king and queen. The Khan invited them all to sit down as a male servant poured wine into each of their cups.

"A few weeks ago, I was in a hunting party, and we ran into Hatcher and Joshua. Over the next couple of weeks, we ran into each other a couple of more times. On each occasion, it seemed to be either lunch or dinner, and we shared meals together. I got to know Hatcher and Joshua. The second time we met, Joshua let slip that Terra and Hatcher were married. I did not know who Terra's husband was and did not try to find out, knowing that this might be uncomfortable for you, Terra. Now I know who your husband is. I can see that Hatcher is a remarkable man, which is what I could assume without knowing since he won your heart, my love," Ammon said as he turned to Terra, who was hanging on every word the Khan said.

"You Earth people do not share many of our traditions and customs regarding marriage and family. Hatcher knows, from Joshua, that it is acceptable for a man to have more than one wife, should his first wife grant it. This is also a custom among my people. We have yet another custom that a husband can agree to give his wife another man, if she desires it," the Khan said as every face in the room remained fixed on the Khan, who smiled more broadly, then continued, "Terra, I would like to let you also have Hatcher as your husband. Do you consent to this? Hatcher, do you consent to take Terra as a wife? Taylor, will you allow Hatcher to also have Terra as his wife?"

No one spoke for a moment, then Hatcher let out a deep breath and said, "My Khan, sure, this all sounds good—, to me, I mean."

Taylor and Terra started laughing at Hatcher's words.

"So you all consent?" the Khan asked, hoping this issue was now agreed on.

"Yes!" Terra and Taylor answered in unison.

"Good, then we can eat. I'm starving!" the Khan answered.

"Wait!" Taylor shouted, which suddenly stifled the mood. "Do I get you, my Khan, as another husband? Also, Terra and I, well, we have a thing for each other, is that all good to you?" Taylor said as everyone sat frozen in their seats.

"I'm good with that," Hatcher said, smiling at Taylor. "I think this is acceptable to me," the Khan answered.

"One more thing," the Khan said and paused before continuing, "Is she, Taylor, always so, uh, 'aggressive' in vocalizing her desires?" the Khan asked seriously.

"Oh yes, my Khan she'll tell you exactly what she thinks. You'll grow to love it, and her," Hatcher answered with a smile.

Everyone laughed at this, except Taylor, who said, "You men are still not accepting of our position as women! We think for ourselves!" she said.

Everyone laughed again and began eating.

CHAPTER 7

News from the Great Plains

IT CREATED A stir on the first day of November when a single boat appeared in the protected harbor next to the encampment of the Riders. The boat had six people. The leader introduced himself as Awas, a trader from the Great Plains, the home of the Riders. Ammon Khan listened as the man introduced himself and those who came with him.

"Awas, what news have you of my father Elwayne's kin? Surely you know of his kin? My brothers, Jawalen and Sinter, are ruling together in my absence. How goes the affairs of the Riders of the Plains?" Ammon asked.

"Your kinsmen have had great tribulation, oh Great Khan. A sickness has befallen the people; many have died. As if this were not enough misery, the gods have seen fit to send more. The people to the east of the Great Plains are moving more and more people across the Great Mountains. They encroach upon your dominion and have sacked many of your villages. Their leader calls himself after a great leader on a faraway world—the world known by the men who rebelled against The Mothers."

Ammon Khan huffed and laughed. "And what does he call himself?"

"He also calls himself Khan, as you do. His name is Genghis Khan. He says this was the name of the greatest Khan of the world, known as Earth. He means to take all of your empire, oh Great Khan. Your brothers plan to attack him next spring when the winter snows melt, though I fear they will fail, unless you and many of your people return to assist them."

Ammon stood silently upon hearing this news. "Summon the Oracle, my wife. I must consult with her."

* * *

"Can you trust this man, what he says?" Terra asked Ammon.

"I do not know him. 'Trust' is impossible in this case. I need you to see his mind and tell me if what he says is true. I want to invite him and his men to dinner and have you listen to their thoughts."

"I will do as you ask," Terra said.

* * *

Dinner consisted of roast duck and a number of vegetables that were now being grown in one of several large plots of ground on the outskirts of the Riders encampment. As they sat down, the Khan introduced his wife, Terra, who was seated next to him.

"Awas, Ammon tells me you have warned him of a threat from a man called Genghis Khan," Terra said. "What do you know of this man? Is he a man that can be reasoned with?"

"My dear, you need not concern yourself with the matters of men. Let your thoughts remain with hearth and home," Awas said.

"Master Awas, my wife Terra is a Sky Lord, and The Mother of this world. She traveled across the night sky to reach our world and help us as we progress into the formation of empires. She is learned in ways we cannot even conceive of. You should consider her at least my equal," Ammon said, staring directly at Awas.

"I meant no slight, oh Great Khan," Awas said, then continued. "Genghis Khan is a man of great ambition and great resources. His people are expert horsemen, like the Riders of the Great Plains. They have also learned about gunpowder and have made guns based on the learning of The Mother's men who rebelled. I fear it may not be possible to defeat him. His people are more numerous than the sand grains upon this beach outside your encampment. I suggest negotiation. Perhaps he will let you rule your people as a vassal to him."

"No, we will not negotiate. We will defeat him in battle. If necessary, he can then ask for negotiations with us. Our terms at that point will be simple. They must return across the mighty mountains and deliver their Khan to me for his execution. It will be quick and painless though. We decapitate those who attack us.

Let's enjoy our meal. It is roast duck, but you may have guessed that from the shape of the main dish. Perhaps it is more difficult to guess since that guest also lost his head!" The Khan said, laughing heartily.

* * *

"What did you learn of his true intentions, Terra, my love?" Ammon asked.

"He is beholden by fear to the other Khan. His family is in jeopardy. Genghis Khan has threatened his family if he does not return with an agreement to negotiate. I do not know of a way, yet, whereby we may avoid the confrontation this other Khan seeks. But I shall keep thinking. What are your plans, my love?"

"He expects me, I suppose, to attack, man to man, army to army. This I will not do. I know this Khan has a great numerical advantage. I must ask, can you plant ideas into the heads of people, even if the ideas are false?" the Khan asked.

"I can. What do you have in mind?" Terra asked.

"It is but an idea at the moment," Ammon said. "I must think it through more."

"Let me know when you are ready," Terra said.

CHAPTER 8

The War Council

"I AM ASKING IF I can borrow a dozen of your stun guns and the shuttle. I cannot say when I shall return with it. You will need to give me at least two pilots and a couple of your security people who know how to use your rail gun," Ammon Khan said as he looked at President Fisher Davis and the assembled congresspeople.

"We would need to have Terra approve such a risky plan, Ammon. Why is she not here with you to present this idea?" President Davis asked.

Ammon paused, looked over the room again, and said, "She is home with her children. She is working to train her daughter, Constance, to take her place while she is gone with me and my men."

"Constance is but eight years old! Can she master the gifts enough to provide us with what Terra does?" Carla Chan, one of the congresspersons, asked.

"I can—at least enough to tell my mother of any problems," Connie answered as she stood next to her mother. They had just entered the room, unnoticed by many of those assembled.

"In my absence, Connie will be able to communicate with me at any time and provide the support this colony needs," Terra answered.

"Do you approve of this plan, Terra?" President Davis asked.

"I do, for the simple reason that this is a threat not just to the people of the Riders, but also to us here as well. This man, Genghis Khan, has unchecked ambition and enough resources to accomplish any aims he has. His people have gunpowder and gun technology. If they conquer the people of the Riders, they will force them to build ships and sail to our shores to fight and defeat us. He sees us as his only real threat. He must be removed from power or his mind changed. Doing this will mitigate much of the bloodthirsty ambitions these people have. Ultimately, we must do for those people what we have done for Ammon's people here. We must teach them of more peaceful ways. This is not a threat to ignore," Terra finished as she looked over the room, into the eyes and hearts of all those assembled.

President Davis was silent as Terra finished speaking, then he spoke, "We shall put this to a vote. All those in favor of the Khan's plan raise your hands."

The vote was unanimous in support of the plan.

CHAPTER 9

The Journey Across the Great Sea

"I NEED HATCHER AND Gunner. I know this worries you, my dear, but Hatcher is a great pilot and Gunner has a warrior's heart and knows how to employ your weapons to our advantage," Ammon pleaded with Terra to allow Hatcher and Gunner to go along with them.

"My children will be without their fathers and their mother. This leaves so much responsibility on Taylor," Terra said as she turned and stared at Ammon and started to cry.

"My love, your parents and Hatcher's are there to help out. They will be well cared for. Your presence on our mission is an effort on your part to protect them from threats they do not even see yet."

* * *

"Terra, I am a Terran. We are strong people. I can care for the children. You must go and protect us, it is part of your duties," Taylor explained.

Hearing Taylor's words seemed to help Terra feel better about leaving her children in her care.

"I consent. Taylor, I love you. I know you will watch over the children just as I would do," Terra said as she went to Taylor, who hugged her deeply while lifting her off the ground.

* * *

Two days later, the packing of the shuttle was complete. It was fully loaded with 120 Riders and a number of Earth people as well as many supplies they would need for a trip of an undetermined duration.

"Benston Control, how do you read me?" Hatcher asked into the microphone on the control panel on the upper level of the shuttle.

"We read you five by five. Good luck, Hatcher, Godspeed," Sam replied from the com room in the center of the Benston Colony.

The spinner blades began moving slowly then gained speed as the thorium-powered engine reached 80 percent power. The shuttle lifted slowly, straight up, and then it began moving northeast out across the Terran Gulf. Their destination was the shore of the Great Plains, seven hours distant at top speed. The rest point for day one that was selected was the location for expedition number three, which was never attempted after the events of expedition number one, some years earlier.

Everyone was excited. For the Riders on board, who constituted most of the passengers, they were going home. For the colonists, they were seeing a part of the new world that none of them had seen before. For the first two hours they crossed an arm of the Terran Gulf that was unnamed at that time. Then they crossed over a smaller peninsula of the Iberian Peninsula that was also unnamed. Lastly, they crossed back over the Terran Gulf for three more hours, and then the coast came into view again. This part of their new world was an entirely new continent. Those on board wanted to continue naming land masses and seas after places on Earth. Every one of the colonists on board wanted to call this new continent Asia.

Hatcher had picked out a landing spot from the maps developed by Peter when the *Calypso* mapped the entire world from orbit, some years earlier. Even with the maps—and a good idea where some flat areas were located—Hatcher still had to make a final decision on a landing spot for the first leg of their journey.

"I guess we are both a little nervous after what happened last time, huh?" Gunner asked from his copilot seat.

"Yeah, well I did pack two descent parachutes and the liquid fuel tanks are full, just in case!" Hatcher said with a chuckle.

"Checked over the gearbox assembly too, I suppose?" Gunner asked.

"Yup, inspected it myself, though everyone on the maintenance team was pretty careful this time around," Hatcher said. "Probably the best thing that can happen on a complicated flying machine, you know—have an accident that everyone survives, but one where everyone learns good lessons."

As the shuttle crossed over land for the first time in several hours, Hatcher began looking for a landing location in the vicinity of the ones he had selected before leaving. The land was drier here—the outside temperature was nearly 38 degrees Celsius (100 degrees Fahrenheit) in the late afternoon. Hatcher found a large flat area to land and brought the shuttle down for the night.

Gunner had Terran guards set around the shuttle overnight. He also ejected two drones to circle overhead to alert them to any hominid activity during the night.

Gunner spoke to the Earthers on board. "All right, everyone, the six Terran guards will take the first watch, then I want another group of six to take the second watch of the night. Seems fair that it should be Earthers on the next watch. We all need to get some rest. Any volunteers?"

Six Earthers volunteered, but Ammon said six rested Terrans should take the next watch since they knew more about the land and peoples here. Gunner agreed. His logic was sound.

* * *

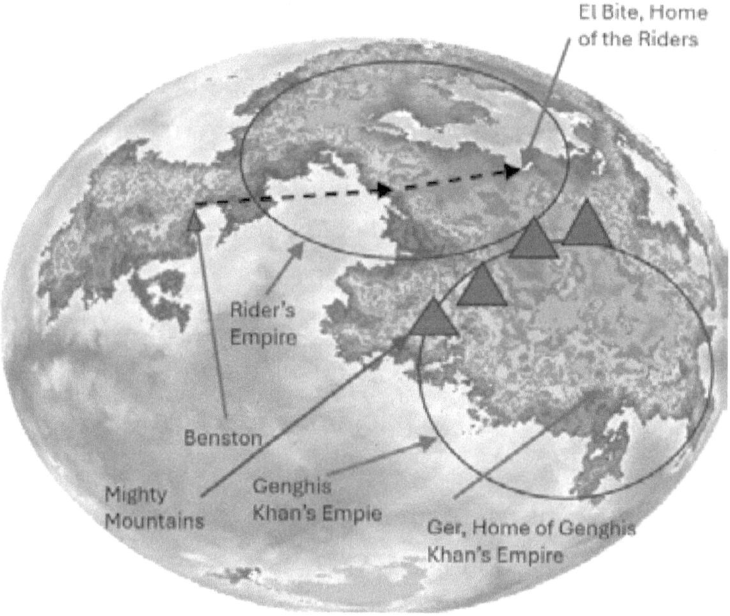

The Western Hemisphere of Terra Nova

CHAPTER 10

What Has Befallen Our Home?

AT DAWN THE next day, the cooks were up early, passing out ready-made breakfasts. For the Terrans, this still consisted mostly of meat that had been salted away for winter. Terra and the nutritionists were trying to change the diet to a more diversified one, but the change was slow in coming.

On the flight deck, Hatcher, Terra, Gunner, Ammon, and two of his men were discussing the route of travel and what they could expect upon reaching the Riders' capital, El Bite.

"Do your brothers know of Terra, and what do they know of 'us'?" Gunner asked.

"We had been busy raiding before we met the Sky Lords. Once we met your people, we had to travel to your colony, and we began immediately to prepare for winter. We have not sent any messengers. They won't know a thing about 'all of this,'" Ammon said, motioning to the ship and all that it entailed and meant. "It will be a lot for them to accept. Perhaps there is a better way to help my people accept your people. What if we traveled closer, landed, left some people here with the shuttle, then traveled overland the final several days? They would only see you as fellow travelers—a part of this world. I suggest this." They all agreed that this was wise.

That day, they landed about a three-day walk from El Bite. Terra, Hatcher, Gunner, and twelve other Earthers set out with all but six of the Terrans and the remainder of the Earthers. The people remaining with the shuttle would be in radio contact with those preceding on foot. The shuttle was to be guarded from any attacks. To that end, Peter Chan, who had come along, was to fly the shuttle to safety if need be.

"I don't have nearly the flight hours that you and Gunner have in this fickle and complicated flying machine, but I believe I could do what is required to get it to safety if needed. I hope it is not needed. I've never been in the pilot seat, just the copilot position," Peter said, clearly less than 100 percent confident in his ability to fly the shuttle.

"Well, you have the AI instructor, and you could try letting the AI fly if required, though it is always a slow process to let it fly. It never skips steps, and sometimes emergencies require more speedy takeoffs. But, hey, chances are less than 50 percent that you will have a need to fly the shuttle due to an emergency! Just keep practicing with the flight simulator while you are here. We all learned that way in the beginning," Gunner said with a grin.

* * *

They had no horses. No pack animals and nothing else except stun guns and their curious clothing that would give away that they were in any way different than anyone else on this high, hot, dusty, and desolate plain.

"It's so hot and *dry*!" Gunner said as he trudged along with Ammon and his men. "You're sure about the locations of these natural springs?"

"Yes, I am sure. It is but a few more hours. We are going to be fine. We have water for three days, and the springs are something we will reach later today," Ammon said, looking at Gunner with a little annoyance in his glare.

Hatcher was walking with Terra, who was one of only two women in the group. Jessica Whitehorse was the other woman—a member of the colony security force.

"My ancestors were Cheyenne Indians in Montana and Wyoming. Allegedly, I have their genes, so trudging along through a high-desert plain is in my blood. It just doesn't feel like it!" Jessica said with a slight laugh, hoping it hid her fatigue from the heat.

"Well, I have no such genetic advantage!" Terra answered as she stopped momentarily to wipe the sweat and dust from her face and took a drink from the water bladder that was on her back.

"Can we take a break? The women are exhausted. Those large boulders up ahead seem to have created some shade," Hatcher said.

One of the Terrans grunted and said, "We'll rest when we reach the springs. There is shade there and water, of course."

Ammon turned toward his man who had just spoken and said, "We can rest for a few minutes by the boulders."

Terra whispered to Hatcher, "Thank you. I don't know how much longer I could have gone."

The shade was only large enough for about half of the group to find rest in, so they took turns. This particular day was very hot, even the Terrans made comments about the heat.

As they rested and drank water, Elias Whitcomb, who was with Hatcher when they were captured by the Terrans a few years earlier, was looking through his binoculars into the distance.

"I see somebody. Somebody is coming toward us. He is on horseback."

"Let me look," one of the Terrans asked. He took the binoculars and said, "He is one of our people."

Ammon and two Terrans spoke in their native tongue with each other. Then began walking toward the man. They met a couple of hundred meters from the boulders and then stopped and began conversing. Their conversation could not be heard by anyone else. After only a moment, they all began walking back to the boulders.

Ammon spoke to the group, "This man was a servant in my father's household, his name is Darsam. I know him. He says El Bite has been sacked by the other Khan, Genghis Khan. This was not an action taken by his people intruding into my lands. This was an act of war. He has continued onward, attempting to sack other villages. We must reassess our plan."

CHAPTER 11

Intelligence Gathering

THEY STAYED THAT night next to the boulders. Gunner had suggested that the long-range reconnaissance drones from the shuttle may help locate the armies of Genghis Khan, or GK, as they were now referring to him. They radioed Peter at the shuttle, who ejected three drones to different villages in the general vicinity of El Bite.

The next morning, all three drones were circling high above the three villages. The com radios they brought with them had fold-out screens, which allowed them to see the video feeds from each drone. Careful scrutiny of the video showed that one of the villages had a very large group of horses and men that did not look like Ammon's people. It appeared that the new horsemen, presumably men of GK, were camped at that village, possibly intending to rest there for a few days.

"If they behave similarly to your men, they will be sending out patrols into the countryside to assess any threats to their troops. Perhaps we can capture some of these men and learn of their plans," Gunner suggested.

"Yeah, sure, because it worked so well for us the other time we tried it!" Elias said, not laughing at all.

"It was a decent plan. We can modify it—improve our chances this time," Hatcher said somewhat sheepishly.

"It is a good plan. It is what we would do. With your drones, we can know where they are, though we have no horses. We need a dozen horses to make this work, and we have none," Ammon said, dejected.

"My Khan, my kinsman is but a day's journey to the west. He has a great many horses, and some men also. He would be pleased to help you. Gods know it! We need help against these marauders!" Darsam said.

"Let us break camp and march with haste to your kinsman," Ammon said.

* * *

By early evening, after a grueling march that left everyone exhausted, they reached the top of a mesa where Darsam's kinsman, El Awell, dwelt in a collection of two dozen tents, with over a hundred horses grazing lazily on the lusher top of the mesa. The temperature seemed noticeably cooler on the mesa top. A small spring provided water that was directed into shallow ponds for the use of the horses and people.

Darsam and El Awell spoke while Ammon and his men waited, listening to the conversation. The Earthers, with the exception of Elias, could not speak or understand this Terran dialect. Elias had learned some of it from his time hunting with members of Ammon's Riders. He translated for the group.

"He is willing to help in any way. GK threatened his family and their horses. Some of them were killed when El Bite was sacked. He is more than happy to get revenge."

* * *

"Our drone shows that patrols are sent out in the four compass directions every morning. They travel about three hours out and then come back. Another patrol is sent out in the afternoon, and those patrols stay out all night, then begin the return leg before dawn breaks. This patrol is our opportunity. We can ambush them while they sleep. I propose a two-pronged approach. Our primary attack force will seek to surprise them in the dark. We will use our night-vision tech to gain an advantage over them. Should any of them escape, we will position a second group on the route back to the main encampment. This will prevent GK's men from learning what happened to them," Gunner explained to Ammon's men, Hatcher, Elias, and Terra.

"We'll take six of our men," Hatcher explained. "Four will be with the main attack group. They will use the stun guns on wide settings to stun as many as possible. We want to minimize loss of life. Two more with stun guns will go with the second group to stun any escapees. Again, lethal force needs to be a last resort. Killing will only inflame an already bad situation."

At midnight, the attack groups left the mesa and headed west to the closest of the four overnight patrols. It took three and one-half hours to reach a point where they were close to the sleeping enemy. Gunner and his three Earthers with stun guns crept near the camp and prepared a coordinated use of the stun guns on the sleeping enemy. As they crept forward, using their night-vision tech, they noticed a sentry positioned on the perimeter of the camp.

"Elias," Gunner whispered, "Take the sentry out. The rest of us will take the sleepers."

"Copy that," Elias whispered back.

A whoosh and a modest pop of static electricity sounded as the sentry was hit and dropped to the ground.

"Shit, the sleepers are going to wake up!" Gunner whispered to Hatcher as they spread themselves out to fire their stun shots at the sleepers.

Three pulses of stuns created even more sound as sleepers were hit by the static fields.

"No one is moving. Did we get them all?" Hatcher asked Gunner.

Each of the four lit road flares so they could look at the group of stunned sleeping enemies. No one was moving, but that would not last long. Gunner whistled for the Terrans to come in and zip-tie the stunned enemy. As they were doing so, one stunned enemy soldier stood up and ran toward the horses. Elias fired, but the distance was too great to stun the running man. He got on a horse and was riding away. One of the Terrans took one of the horses and pursued him.

Hatcher fired an ariel flare to illuminate the surrounding terrain, hoping it would help the Terran catch the fleeing rider.

"No, no, no flares in the sky!" Gunner shouted. "There may be people who will notice that."

"Sorry, I—" Hatcher started to say.

"It's OK. We'll deal with whatever comes of it," Gunner shouted to those in the group.

The stunned men were rounded up and taken back several hundred meters, where Ammon, Terra, and three Terrans waited.

Terra walked up to the first restrained man and offered water to him. She helped him drink it since his hands were restrained.

The man looked at Terra in the weird red light produced by a lit road flare. His tenseness seemed to ease and he nodded to her, an indication that Terra took to mean "Thank you."

Terra pushed a thought into the man's head. The thought was in the form of an image with Terra saying, *My name is Terra. What is your name?*

The man smiled again and thought back to her, *Zhe.*

"Zhe," Terra smiled and thought back to him with an image of a house and asked in thought, *Where is your home?* The man was from Ger, the capital of Genghis Khan's empire. The interrogation lasted for four hours.

The sun was now up. The patrol would be missed soon. They needed to finish up and ride away.

They learned that GK was not with the large group but was with another group even farther west, getting ready to surprise attack three more villages that day.

"We have to warn those villages, but there is no way we could get to those villages soon enough. Also, we would likely be captured by this large force were we to try," Ammon said, now desperate for options to save his people and stall GK's drive westward.

"If we want to stop this attack and turn it back, we will need to use the shuttle to get there, and we will need to use the rail guns to stop them," Hatcher said.

"We can't do that. Many would die," Terra said fervently.

"We could fire warning shots in front of the advancing horde of attacking horsemen," Gunner said.

"That might work. Just seeing the shuttle may startle the attackers into a retreat. Let's assume that it may work. What is the larger plan? GK will be in this group. Ammon, what is your plan after this?" Hatcher asked.

"I want him to know who Terra is and what she can do. I want him to be seduced by Terra," Ammon said.

"Now hold on, Ammon. Terra is not going to be used in that way," Hatcher shot back.

"I don't want that either, Hatcher. This would only go as far as making him think she might like him as well. Once he believes that, Terra can make him think we are invincible and that good relations and trade between our empires are the best option for him," Ammon explained.

"OK, let's get on the radio," Gunner said. "We'll have the shuttle come to our position, pick us up, and reposition in front of their largest force. At that point, there is going to have to be some ad-libbing. We'll have to be ready for whatever happens," Gunner said.

CHAPTER 12

Confrontation on the High Plains

PETER ENDED UP deciding he needed to fly the shuttle himself. He was, after all, an engineer, and engineers don't actually always trust technology, strange as that may seem. He lifted off and was headed to the pegged location of Ammon's group, but he was sweating through his shirt by the time he landed an hour later, despite having the AC vent blowing 18-degree Celsius air on him at the max fan setting. By then it was midday and everyone on the ground was happy to be back inside an air-conditioned compartment.

Hatcher had Peter move to the copilot seat as he lifted off to head to where they believed the men of GK would be. Gunner was needed in the weapons pod on the bottom of the shuttle.

"Goddamn it! This compartment is a tight fit! It reminds me of the ball turret on the B-17s I saw in the museum as a kid in Mesa, Arizona," Gunner shouted as he tried to fit into the cramped space. "This thing must have been made for Hobbits! Not for a 193-centimeter tall, 109-kilogram African American male!"

"Whoa, I think I see his men! Look to our two o'clock position—lots of horses and men. That's gotta be them!" Peter shouted through the com.

"Yo, Pete, settle down! I think you broke my eardrums! I see them. Let me fly the shuttle through the controls down here. We need to get in front of them. I'm sure they see us, even though we are at six thousand meters altitude." Gunner said back to Peter through the com system. "OK, I've got us about a thousand meters in front of them. I'm going to fire a three-second burst, see what they think of that."

The three-second burst was sixty rounds of depleted uranium shells. These supersonic shells hit the hard surface of the ground, causing an eruption of rock and dirt in every direction. Many horses were startled and threw dozens of Riders off as the horses galloped away from the danger they perceived. The entire group of horsemen stopped, and many turned and fled.

"I need a Terran on the outside PA speaker, stat. I want this in English and the Rider's dialect," Gunner shouted.

"I'm on it!" Ammon shouted as Peter handed him the microphone.

"Men of Genghis Khan, dismount your horses, throw down your weapons, or you will die!" Ammon said in both languages two times each.

Gunner fired two more three-second bursts on each side of the fleeing horsemen. This had the intended result as hundreds of horsemen dismounted and dropped their weapons.

By now, men from the village were arriving. Peter put the shuttle down, and every one of Ammon's men rushed out and began rounding up the disarmed men, while the Earthers began picking up the weapons on the ground so that they could not be picked up and used against Ammon's men or the villagers. Earthers with stun guns came out of the shuttle and pointed their weapons at groups of GK's men.

"Where is Genghis Khan? Identify yourself!" Ammon shouted as he walked among the prisoners.

No one stepped forward for several minutes. At last, a man stepped forward and said, "This is him. This is him!"

The man who was identified as GK stepped forward to Ammon and said, "I am he whom you seek."

Genghis Khan was tall, skinny, and had a very Asian look to him. He also looked very young, probably no more than eighteen years old.

"I ask you to have mercy on my men. You may take your vengeance on me, but I beg you to let my men return to their homes. The actions against your people were entirely my doing. I succumbed to the sophistries of my soothsayer, who convinced me that I would easily conquer your people in your absence, oh great Khan Ammon."

CHAPTER 13

Can We Trust His Word?

AMMON WAS MEETING with his brothers, Jawalen and Sinter, and Terra concerning the plan for getting the rest of GK's men to surrender. There was also the issue of what to do with the other Khan.

"He should die. I recommend death by venom—black scorpion," Sinter said.

"I agree," Jawalen added.

Terra frowned and said, "Killing this young man will turn him into a martyr. His people will always have revenge on their minds. If we spare him, he could become an ally."

"But how do we know he will keep his word? I fear that by letting him live, we may end up fighting the same enemy again," Ammon noted.

"Let me interview him, my love. I can ascertain his intentions." Terra turned and left the flight deck and went below to where the young Khan was being held.

"How are you doing? Is there anything I can get you?" Terra asked Genghis.

"I require nothing but thank you for asking. Your treatment of me and my people is far better than I expected and much better than I would have treated your people if the situation were reversed. I must ask, are you what the men who rebelled against The Mothers called an 'oracle'? Does this mean my thoughts are known to you?"

"Yes, I am an oracle. I have been anointed The Mother of this world by the great Mother of my world, who is known as Eve of Eden, and yes, I can see your thoughts."

"Are you here to prepare me for my death? I assume it will be slow and painful. I do not look forward to it. I had hopes of having a family of my own someday, but my pride and vanity have made this impossible."

Terra saw the sadness in the young man's eyes and felt it in his thoughts. She felt great sadness for him.

"The men upstairs are trying to decide your fate, and yes, they want you to die. I have argued that this is not the correct course of action. We want you to reform yourself and become an ally. There is no reason for any more to die. Great advancements for both peoples are possible by building friendships and trade between these two great empires. Can you accept this course of action?"

"Why do you ask me? You can see my thoughts. I cannot lie to you. I can accept peaceful relations between our peoples, but you know that already. My heart has been changed as I contemplate my death and the loss of my future. In another time and place, perhaps you and I could become friends, perhaps lovers. I have no woman."

Terra smiled at him and held his hand. "I will tell them you have every intention of pursuing peace from this time forward. They will listen to and accept my opinion because of my position."

"Thank you, Terra. I am in your debt. Someday you should visit my empire. There is much I could show you. We are an industrious people and have many beautiful things that a woman of your stature would enjoy. I am so interested in learning all of the things you know. I am told you traveled a great distance across the night sky to help the people of this world."

"Yes, I left my world at ten years of age. After we left, my world was destroyed by the same men who rebelled against our Mother. Also, yes, I would like to visit your home. I will now go back upstairs to speak to the men." She squeezed his hand and smiled at him again. He smiled back as his gaze lingered upon her as she walked away.

CHAPTER 14

Returning the Other Khan

JAWALEN AND SINTER did not like Ammon's decision to spare GK's life. They also did not like the influence Terra had over Ammon.

Terra spoke privately to Hatcher. "I am worried that Ammon's brothers may still seek revenge against GK. To protect his life, as I promised him I would, we need to fly him back to Ger. I know Ammon wants to stay a while here with his brothers and visit his father's tomb. Can we return him? Do we have the supplies and time to do this, Hatcher?"

"Yes, of course, we have the time and supplies. We planned for three months. We've been here only a month. Many of the Terrans who came with us want to stay here. I'm sure Ammon will support your decision to return GK."

* * *

Four days later, Hatcher and Gunner lifted off in the shuttle with all of the Earthers, who were more than happy to leave this hot, dusty place. The flight plan was to fly directly, in one day, to Ger with GK and a number of his top commanders. They left before the sun came up because of the length of travel required that day. The route of travel would take them over the Mighty Mountains, which were the tallest mountains on this world, the heights of which exceeded Mount Everest on Earth.

Terra remained below with GK, his men, and many of the Earthers. The flight was smooth in the cool early-morning air, but that changed as they climbed over the Mighty Mountains. By then, the hot air from the desolate plain was pushing up against the tall mountains, causing the shuttle to experience the roughest ride that any of them remembered aboard the shuttle.

"Everyone, remain seated with your harness securely fastened. I don't know when things will smooth out. Sorry for the rough ride," Hatcher said over the intercom.

GK and his men were not tolerating the flight well. A number of them became nauseated and threw up into the barf bags in the seat backs. About two hours later the air became cooler. The video monitors showed snow and ice on the landscape below in all directions. The cooler air made the flight smoother and Hatcher told them they could get up to use the restrooms.

The restrooms were a new experience for the young Khan and his men. They were used to relieving themselves wherever required, which was not appropriate in the shuttle. Jessica Whitehorse had to show each man who tried to use the bathroom how it worked.

By noon, everyone was hungry. Terra and several of the women began handing out ready-made meals, which was, again, a new experience for these Terrans. The meals had fruit, boiled eggs, and some salted deer meat. The drinks that were available were tea and water. It turned out that tea was known to these people, who seemed to enjoy it.

An hour after lunch, Terra found a seat next to GK and sat down beside him. He seemed greatly pleased that she sat by him. Terra noticed he was not comfortable with flying. He seemed genuinely frightened of being so high up in the air.

"I suppose this experience is not new to you, Terra? I mean, you traveled many days across the night sky to reach our world," GK asked in a shaky voice that failed to hide his fear.

Terra smiled and answered, "Flying through space is much different than flying in the air, as we are now. It is not bumpy. It is, in fact...well, quite boring, and very quiet."

GK seemed to become more relaxed as he listened to Terra and noted that she was not fearful in any way.

"What is your world like? I suppose it is more—what is the word I want to use—complex?"

"Maybe you mean 'modern'?" Terra asked.

"Yes, I think so. My English—that is your language, right? My English is not so good. My fathers learned your language from the men who came long ago, but much time has passed, and I fear we are changing your language."

"Your English is quite good. Much better than my poor ability to speak a Terran dialect," Terra said.

GK smiled at the compliment.

"GK may I call you that?" Terra asked. "Is your father alive? I mean, you are so young to be a leader of your people. On my world, leaders are usually older, sometimes quite old when they begin their reign."

"My father died when I was four years old. My mother died when I was ten years old. My mother's sister raised me after my mother died. I only began my reign a year ago, taking over from the regent who ruled our empire when I was too young to do so."

"GK, what is it that you want to accomplish during the time you are the leader of your people?"

"I desire my people and my kingdom to prosper and become respected in this world, but I am not sure how I will do that now."

"Why do you say that?

"Because I failed in my quest to conquer the Empire of the Plainsmen, or—as you call them—the Riders. My people will not respect me now, though I see now it was foolishness on my part. We can gain so much more through peaceful pursuits and cooperation."

"We all fail. It was said, once, by a wise man on my planet, 'Show me a man who has never failed, and I will show you a man who never tried.' We live, we fail sometimes, we succeed sometimes, but we always learn and improve. In fact, that is a basic tenet of my mystic beliefs. We are on an eternal journey to experience and grow as we live each of our lives."

"Lives? Maybe I do not understand. Are you saying we live more than one life?"

"Yes, we do. We could never learn and experience everything in one life or several. It takes many lives to reach a point where we perfect ourselves. I am still working through many challenges in this life. Sometimes I succeed, sometimes I fail, but it is my hope I will be better, closer to perfection, at the end of my experience here."

"I find this interesting and compelling. Our mystic beliefs are not so well explained. Our shamans tell us we must accept the Great Spirit into our lives or we will suffer for eternity after we die."

"My world had many of the same teachings, and they were all wrong. They were created by people, used to control other people, and gather power over large groups of people. Eventually, we rejected these ideas, and at that point, we became so much happier as we realized The Ones fully intended us to not become discouraged; we would have the time needed to perfect ourselves through living and learning in each life we lived."

"I would like you to teach this to my court personnel, and then I want my people to learn of this."

"I will do this for you, Genghis Khan. Your heart is pure and open. You will progress so very far in your life because of this."

CHAPTER 15

The Empire of GK

IT WAS LATE in the evening when they arrived in Ger, and this was a bit of a problem.

"Where can we land? Will it be a problem for your people to see the shuttle?" Hatcher asked GK, who had been brought to the flight deck as they approached Ger.

"The Royal Gardens are very large, with several areas that have not been cultivated. Landing there will be possible, because you need a large flat area to land, correct?" GK asked.

"Yes, large and flat, and it would be best if it were out of sight of large numbers of people. It is a lot to ask of people to accept that someone has come from 'across the night sky.' Easing into that idea is probably best," Terra said, and Gunner and Hatcher nodded in agreement.

By this time in the evening, it was almost dark, which was making it more difficult to assess a landing site. Using night vision helped, and within thirty minutes of circling and looking for the Royal Gardens they located it and were able to land.

As Terra, Hatcher, Gunner, GK and his men walked down the ramp they were met by heavily armed men on horseback. GK went up to the men and spoke to them. This seemed to soothe the men, who became visibly less alert. Apparently, having their Khan tell them that these very unusual strangers were friends satisfied their concerns.

"Terra, Hatcher, Gunner, come to the palace. There are quarters for you there. I know that the rest of your people want to remain here with the shuttle to protect it. I understand that. I will have my guards keep people away from it. Perhaps tomorrow, many of them can also come to the palace. It is late. Let us make the journey to the palace," GK said.

The journey was made with two horse- drawn carriages. Terra and the Khan went together in one, and Hatcher and Gunner in the other. The journey took almost thirty minutes. When they arrived, palace personnel took Gunner and Hatcher one direction and Terra and the Khan another. Hatcher was a bit concerned.

"I hope he doesn't try anything. We have deliberately not told him that Terra is married and has children," Hatcher whispered to Gunner.

"He seems to be a gentleman. I think she'll be OK," Gunner whispered back.

* * *

"I hope you find these quarters to your liking, Terra," GK said when they reached the door of a large bedroom. "My quarters are just down at the end of this hall. My staff will be stationed outside your door, should you need anything."

"The accommodations look like what we call a five-star hotel room on my planet!" Terra said, smiling as she did.

The Khan smiled back and said, "I will see you in the morning at breakfast, Terra."

* * *

In the morning, Terra realized that she had brought no clean clothes. Just then she heard a knock at her door.

"My lady, the Khan has asked me to bring you clean clothes," she said as she handed clothing to her. "We will draw your bathwater and help you bathe and dress."

The young woman who had spoken to her and two other women walked in and began dumping containers of water into a large metal basin in the room. One came over to Terra and began removing her clothes. Another one carried them away, saying she would see that they were washed properly.

Terra was brought over to the basin, now full of warm water, and stepped into it and sat down. The two women pulled out sponges, poured a purple fluid onto them, and began to gently wash Terra. The odor was lavender and seemed to have a calming effect on Terra, which she enjoyed. It also made her quite hungry, for some reason.

After being dried off and dressed, Terra's hair was brushed and prepared by piling it up on her head, a style she had never tried. She was led down the hall and descended wide stairs into a great hall with a long table in the middle. The Khan, Hatcher, and Gunner were already sitting together at one end.

As Terra walked, the long purple, pink, and white robes she had been dressed in slid along the floor. One of the women hurried and picked up the portion trailing her on the ground.

The Khan stood, as did Hatcher and Gunner as Terra came down to their end of the table. They were all staring at Terra, mesmerized at her transformation. She was stunning.

The servants brought in breakfast, which was heavy on fruits and juices, with some protein in the form of sausages. At the end, tea was served.

"I would like you all to have a tour of the great garden, the palace and its grounds and then I'd like you to see the city. Does that sound acceptable? I will have the royal palace keeper guide the tour. I myself must attend to my duties here, which have suffered greatly in my extended absence," GK explained.

* * *

"I take you first to the great terraced gardens, built on the mountainside—that mountain over there,"

Sennacherib, their guide, pointed to a mountain several kilometers distant.

The ride there took almost an hour. First, the carriage wound through the cobbled streets and densely packed residences just outside the royal palace. Then they reached the beginning of a wide, long, cobbled road that led straight out to the great terraced gardens built on the mountainside. The road was elevated above flooded patches of paddies. These people had clearly already started developing agriculture.

When they finally stopped, they were met by two dozen men who would carry them in "chairs" to the top of the terraced gardens. The tour would start at the top and proceed downward to where they had stopped.

The climb to the top of the gardens took another hour and fifteen minutes. The men carrying them in the chairs would stop periodically and fresh men would begin anew, carrying them to the top. When they reached the top, the view in the distance was stunning. It was green everywhere.

Their guide took them through each level, showing them the many exotic trees, flowers, and shrubs.

"Sennacherib, why was this beautiful garden built? It is beautiful—very, very, beautiful—but we are the only people here, at least today," Terra asked.

"These gardens built for beautiful princess brought here to be queen, many years ago. She came from land farther south, land that was, what you say? Lush, I think? Khan at that time called a king, it was older dynasty. Beautiful queen cried herself to sleep every night thinking of her beautiful home, far away. King, he sad for his wife and built garden for her."

"Was she happy once he built the garden for her?" Terra asked.

"Yes, she happy. King also bring her family from south to here, that make her happy also! It said, she most beautiful woman in world. Though maybe you now most beautiful woman in world," Sennacherib said, smiling at Terra.

"Well, I doubt that. I've seen the women here, some who are 'just servants,' and they are very beautiful!" Terra said.

* * *

The rest of the tour included the palace grounds and the marketplace, which was both interesting and, well, somewhat nauseating.

"What are the items in the jar?" Gunner asked.

"Monkey brains, monkey balls! Said to make penis bigger, stronger!" Sennacherib said, laughing heartily.

"Does it work?" Gunner pressed.

"How would I know? I eunuch!" Sennacherib laughed again as Hatcher and Gunner exchanged a pained look at each other.

When they returned, it was dinnertime. GK was back and greeted them as they were brought in for dinner.

"Did you enjoy tour?" GK asked the group.

"Yes, the gardens were so beautiful," Terra said. "On our world, in antiquity, in the capital of a great empire,

Babylon, legend has it there were beautiful hanging gardens made for a princess who missed her homeland. It was one of the seven wonders of the ancient Earth. The story we heard today seemed so similar. I believe, if I remember correctly, that there is no remaining evidence of the gardens to authenticate its reality, though several great historians of the time corroborated its existence."

"I liked the marketplace. So very interesting, especially the monkey brains and monkey balls!" Gunner said with a hearty laugh.

"You think you might like to try that?" GK asked with a sly smile.

"Sure," Gunner answered, not entirely with enthusiasm.

"Good, it on menu tonight for dinner! Just for us men, though," GK said, his smile growing larger. "You must try it."

Gunner and Hatcher exchanged a worried look.

"Don't worry, I joke," GK said. "I not believe bull crap story about it," GK said, still smiling and laughing. "Is 'bull crap' right word for this?"

"Yes, you nailed it," Hatcher said, now smiling himself at the Khan's use of the earthy term.

"You two, don't corrupt his language skills!" Terra chided Gunner and Hatcher, smiling and laughing herself.

CHAPTER 16

The Great Library of Ger

THE NEXT MORNING at breakfast, the Khan asked what else they would like to see.

"We have an enormous library on the palace grounds. It includes ancient writings left by the men of The Mothers who rebelled. Many of those writings are from the great Mothers, I am told. I have not had the time to read them myself," GK explained. "Perhaps you would like to read some of them, as you are The Mother of this world now."

"Oh, that would be so interesting!" Terra exclaimed.

"Good, Sennacherib will show you the library. Gunner and Hatcher, would you like to see it also? If not, feel free to just look around the city if you prefer. I will send a couple of my guards to go with you, to protect you from those who might be too interested in seeing you. You are the subject of much conversation among the people, I am told, especially Gunner. No one has seen a Nubian prince before. Your size and strength are already much admired. Many women desire you."

"I get that a lot, you know," Gunner smiled and laughed as he said this.

* * *

"This is the library," Sennacherib pointed to three large buildings in the distance. "We can walk there; It is a beautiful day."

"This first building contains the writings of wise men who served previous kings and khans. The second building has writings of other empires, including ones that preceded our empire—the ancient one, or Empire of the Stars. These were the first people who came here, ancient Mothers who, with the men they brought, created intelligent men and women from the original tree dwellers. The last building has the mystic writings of Mothers from other worlds. Much of what is here has never been read. The language they used is unknown to us now," Sennacherib explained.

"I'm not sure where to start. Maybe the second building. I would like to learn about the 'Empire of the Stars,'" Terra said.

Terra was ushered into the second building, which had two large floors. The librarian of this building was a woman named Chan-hu.

"Chan-hu is an expert in all languages and a master of ancient empires and their histories. She will help you with whatever you need."

"Thank you, Sennacherib. I may spend the day reading; perhaps you need not stay with me, especially since Chan-hu can help me, as you said."

Chan-hu smiled at Terra as Sennacherib excused himself, explaining he would return prior to dinnertime to take Terra back to the palace.

"This book contains the earliest history of those who came to this world. It chronicles the journey across the night sky made by the first Sky Lords to come to this world. They came from very far away, though they moved closer to this world, but we do not know where they reside now," Chan-hu explained.

Terra knew where they lived now, and it was very close by, but she knew she could not reveal that to these people. The world of The Mothers was deliberately hidden from the AOW and every other world by deceptive technology.

The world they came from, she wondered, *was that Earth? Did they move from world to world, creating the "intelligent and virtuous species"? As they did so, did they move to a new world of their own each time so they could be close to their handiwork? The work on the Earth had been interrupted by a dangerous comet; would they have stayed nearby if that had not happened? Maybe the world they reside on now was "close enough" to perform their work on the Earth and now on Terra Nova at the same time?*

She began reading. The language was, surprisingly, English, modern English. She learned that the first ship arrived more than two million years earlier. This world already had many species of animals, including very early hominids. The prior four billion years had been a preparation of the world through successive rises and then extinctions of different classes of large animals. "The Ones" had caused, or directed, preparatory work that encompassed billions of years, and then they turned it over to The Mothers for the capstone creation, homo sapiens. She learned this was the pattern they followed in all the inhabited worlds. Terra realized this was also the pattern on Earth.

This world had also suffered, *perhaps deliberately,* from great glacial periods. The last glacial period had ended seven thousand years earlier, allowing the rise and explosive growth of humans. Some empires had formed and Terra knew more were coming. It was a crucial time for this world. On Earth, empires meant an explosion of conflicts and mayhem. That could not be allowed to happen here. It was not necessary as some had postulated when looking at Earth's history. Growth in lives could still come with peace; war was not a necessity for growth as some Earth historians had fallaciously taught.

The information in this library was priceless, Terra thought. *I could spend years reading the written materials*!

Time flew and soon the hour of dinner arrived. Sennacherib came as he promised, and Terra went with him to dinner.

CHAPTER 17

Instructing the Wise Men

FOUR OF THE six advisors to the young Khan were seated in front of Terra in the Khan's reading room. Two other advisors were on official business outside of the capital.

Terra had taught them about the metaphor of the "Wheel of Lives." She was taking questions from the advisors.

"A person born as an idiot is clearly suffering from his poor past life performance. What is our responsibility to help them, given their culpability in their awful situation?" Zse Dong, minister of Tax Collection, asked.

"Minister Dong, every person deserves our respect and compassion. On my planet, we had a saying, 'There but for the grace of God go I.' This means we need to be compassionate and understanding toward those suffering, lest we incur karma and experience the same challenges either in this life or a future one. The tendency toward selfishness and diminished empathy was a central feature in the manifesto of one of the large political parties on my planet up until two hundred years ago. They completely stymied efforts to help others and then passed legislation that only helped the very wealthy! Eventually, we identified the genes that caused these undesirable personality traits and removed them. It was a great blessing in our societies when this occurred. After this, there was broad support for programs that alleviated suffering. I know this pleased The Mothers as they love all of their children."

The young Khan had come in after the start of the discussion and sat, listening intently, saying nothing at all. When the discussion and questions ended just before the dinner hour, he came up to Terra and spoke to her.

"Thank you for sharing these thoughts. These are difficult men to convince. They think too much sometimes and love to argue for no real reason. But they are fair and will accept ideas with merit even if they contradict their own beliefs. It does take time for them to change their opinions. As you discovered, they believe that the status someone is born into is indicative of their goodness. I, of course, know that many people born with great advantages are not always 'good' people."

"We had the same misunderstandings in our societies for many centuries. It was believed that kings and other rulers 'deserved' their status and position which, of course, only came by virtue of their fortunate births and not by any skillful action on their parts. It is easy for people to believe something if that thing justifies their power and wealth."

"Perhaps the teaching of these principles requires many conversations. Do you have missionaries you could leave behind to teach my people these things?"

"No we have no missionaries, and it is important to understand these teaching are not part of a religion or church. They are simply concepts that teach us why we are here and what happens when we die. Other precepts for living happily in our daily lives can come from many sources. There is no monopoly on truth in these beliefs. Many great men and women have developed principles that are worthy of emulation."

"Let us go for dinner. I'm starving!" the young Khan said.

After dinner, GK walked Terra up to her room. "GK, you have been such a wonderful host! We have loved our time here. Your kingdom is so beautiful. I dare say I can't imagine anything more advanced and beautiful on this planet. Would you like to come in and have a drink with me?"

"I would like that very much."

Terra poured him a drink of wine and one for herself. "Please excuse me. I'd like to change into something more comfortable."

"Of course."

A few minutes later, Terra came out wearing a sheer negligee that did not hide any of her features. The Khan looked surprised.

"I came to think you did not see me in this way," the young Khan said.

"You are a most handsome, intelligent, and influential man. I find myself attracted to you!" she answered.

She leaned in and kissed him deeply on the lips while he put his hands onto her shoulders and slid her negligee off. It fell to the ground, leaving her completely nude. He kissed her again on the lips and then on her neck while she moaned softly. He picked her up and carried her over to the bed and knelt over her kissing her again and again on her lips and on her breasts and began his lovemaking in earnest. When they finished, they both fell asleep in each other's arms and slept until the sun came up.

As they dressed, Terra asked if she could return someday with her daughter.

"I would love it if my daughter and I could visit here again. I started reading through the histories of the first Sky Lords who came here two million years ago. I did not get to the part where the Empire of the Stars was created. This is of great interest to me. My daughter is eight years old, so maybe I will wait until she is a little older to visit. She will assume duties similar to mine someday, and it would help her to become more acquainted with the history of Terra Nova."

"So you are married?"

"Yes, I have two husbands, actually. It was not something I expected or wanted at first, but fate intervened and it became clear that I needed to take another husband to accomplish my 'mission' as 'The Mother' of this world. It helps that I loved them both. They are good men, like you! My daughter, whose name is Constance, or Connie, will take a husband someday—probably to someone like you who is a leader of an empire. As 'Mothers' of this world, we are to help guide empires into peaceful paths."

"I would like to meet your daughter. If she is anything like you, she will be a remarkable woman. So you call this world 'Terra Nova'? We always just called this planet 'The World,' but I think the name Terra Nova befits it better," he said.

"I was named Terra Nova, which means 'New Land' in Latin, an older Earth language. I was born on the day my parents received the news that they had been picked to be on the crew of the *Calypso*, our starship that we flew on to come here. We were going to a 'new land' or 'new Earth,' so that is how I got my name."

"It must have been much bigger than your shuttle, which is very big, in my view. How long was your journey here?"

"It was about twelve years, but when we left, we believed it would take more than a thousand years. We discovered some people from a more advanced planet who helped us get here much quicker."

"More advanced than your world? You seem so advanced to me!"

"That is the nature of life. Someone, it seems, is always bigger, better, and more advanced. But of course, that does not always mean 'better.' Sometimes a simpler life can be better and bring us more happiness. Now that I am more awake, I can think of another reason to return," she smiled as he smiled back at her.

"And what would that be?" He smiled even bigger.

"Well, there is always *that* also!" she said, smiling. "But I would like to read more from your great library. I want my daughter to become familiar with the history of this world also."

"You are welcome anytime, dearest Terra."

CHAPTER 18

The Journey Home

A FEW DAYS LATER, the Earthers were ready to head back to El Bite. All of the Earthers ended up getting to see the great gardens and other beautiful locations around Ger. By the time they were ready to go, the shuttle was loaded up with everyone's souvenirs, making the shuttle seem like a cruise ship after a port call in a tourist trap.

"I wasn't able to get Ammon on the radio. Once we reach altitude, we have a good chance of getting him," Hatcher told Terra.

Terra said, "I'd like to talk to him if he picks up. I am worried about him, for some reason."

"Allan Shepherd, I can hear you. Do you hear me?" Ammon asked over the radio.

"Ammon, I read you. How are you?" Terra asked.

"OK. I have a little cold, not feeling great, but nothing serious," Ammon answered.

"We are on the way back," Terra said. "We should be in this evening—late, I am afraid. Take care of yourself. Get some rest, my love."

"Will do." Ammon coughed right after he said that.

"I'm worried about something," Terra said. "That traveler said there was an outbreak that killed many people. We don't have any idea what it is. If some of his men bring it back to Benston and Ammonsberg, we could have the same problems."

"Well, unfortunately," Hatcher said. "We can't do much about it until after we get all the way home."

After crossing the mountains, Ammon called on the radio again. "I think you should know there are quite a few of my men who are sick. I'm afraid it might be what the traveler described. What should we do?"

"Hold on, Ammon. I'm going to get Dr. Sampson up here," Hatcher said, with Terra bent over his shoulder listening in, obviously concerned.

"Ammon, this is Dr. Sampson. Can you describe the symptoms you and your men have?"

"Uh, yeah, I guess we have runny noses, a cough—it seems to be in our chest, not in our throats—and a fever. We are warm. I know you can measure how bad the fever is, but all I can say is we are very warm and very tired. Five of the men who came with us have died. It's bad."

"OK, Ammon, here is what I want you to do. Keep your healthy men away from those who are sick. Wash your hands often. Don't breathe on each other. Try to keep those with a fever well hydrated and cool. Put cold wet rags on them. Heat pots of water over a fire. Let the water boil. Do this in an enclosed space. This might help the chest congestion. Good luck. We'll hurry. I have a few immune booster shots on board. We need to get you all back to Benston ASAP," Dr. Sampson said.

"Can we get Benston on the radio?" Dr. Sampson asked.

"I'll try. Probably not. Once we get over the Terran Gulf, we will be in a line-of-sight situation," Hatcher said.

"We need to set up a triage ward at Benston and isolate it from other patients and the other colonists," Dr. Sampson said. "The first thing we'll do when we get home is to sequence the DNA of this infectious vector. We'll need to know what we are dealing with."

* * *

They arrived at El Bite at 9:00 p.m. The sun was down, but the air was still very dry and hot. For the last two hours prior to arrival at El Bite, they had been hanging plastic across a wide area of the lower deck in hopes this might help prevent the spread of the disease to Earthers and healthy Terrans. Several biocontamination suits were available. Dr. Sampson, Terra, and some other women were dressed in them when they touched down. They had told Ammon to have everybody ready to board once they arrived. They were going to try to expedite their return to Benston.

"We could do a rocket-powered liftoff, put us into a suborbital trajectory, and get to Benston in ninety minutes instead of twelve hours. But that means we'd have to have everyone remain restrained the whole way. If anyone needs medical assistance, it might be impossible to deliver any aid to the afflicted Terrans. What do you think, Gunner?" Hatcher asked.

"It sounds bad, Hatcher. We need to get them back to Benston. We can't do much for them in the shuttle. Dr. Sampson, do you agree?

"Yes. Let's expedite the return," Dr. Sampson answered.

* * *

Dr. Sampson put LIDs (leg inflation devices) on the sick Terrans. They would be pulling several g's on the return, and the afflicted Terrans might not survive without assistance in keeping blood in their chests and heads instead of pooling in their lower extremities. Once finished, she notified Hatcher that everyone was prepared for liftoff.

"Everyone, this is Hatcher. We are going to expedite our return to Benston. This means a scrambled, thrusted, dust off, and ascent. We'll be on a manual-directed, ballistic, suborbital trajectory. You will all need to stay restrained for the entire ninety-minute flight. Use the restroom now, get drinks down now, because it won't be possible after liftoff, Hatcher out."

Hatcher and Gunner had control yokes in front of them. The flight would actually be flown by the pilot, who would fly the shuttle "up and down the line," meaning the pilot would follow the acceleration curve projected in front of them via a holographic projection.

Just before they lifted off, Gunner said, "I'd say this calls for the cowboy hats, don't you think, Hatcher?"

Gunner put a big white cowboy hat on his head, then handed a smaller black one to Hatcher, who nodded, smiled, and put it on.

"Three, two, one, ignition, liftoff," Hatcher said through the com mic.

The shuttle accelerated wildly, soaring high into the night sky, and then disappeared into the blackness of space. After four minutes, the rocket engines consumed the fuel in the ascent tanks. After engine cutout, Hatcher was able to get Benston on the radio and alert them concerning their medical emergency. They would be ready upon arrival back at Benston. What was left in the liquid propellant tank was going to be needed for a controlled descent. There would be no parachute to slow them down, nor would they employ the spinner. All that would take extra time, which they did not have. The descent to the ground at a safe landing velocity would be entirely achieved in four minutes from the rocket engines. It would be a wild ride.

Eighty-six minutes later, the rocket engines ignited again at twelve thousand meters as the shuttle passed Mach seven, glowing brilliantly from the heat of reentry. The speed at landing was forty-five kilometers per hour, not a soft landing. The shuttle actually bounced a meter and a half on landing and required firing the yaw thrusters to keep it upright. The waiting fire crew was certain it would tip over and explode, but it didn't, and

Hatcher would say later he was sure it would not have tipped over. Even so, the thrusters ignited a brush fire that had to be doused.

The drama was so dramatic most of the colonists came out to see what was happening. Some even saw the hard landing and swore they'd never fly on that thing again. Apparently, Hatcher was born two centuries too late. He would have made a great carrier pilot.

After a few short minutes of dousing the fire, the medical teams began evacuating the sick and took them to the triage and isolation area. The healthy Earthers and Terrans were told to go home and self-isolate and monitor for the presence of any symptoms.

Ammon refused to go to the triage area as it was almost immediately overwhelmed with the sick. Instead, he went to his home in Ammonsberg. For the moment, all that could be done was being done.

CHAPTER 19

Outbreak on Terra Nova

"CAN YOU BREATHE in deeply as I hold this on your chest?" Dr. Lynn Sampson asked as she looked at her translator, Taylor.

Taylor translated the question to the child, who had just barely stopped crying. The child's mother looked on with concern and worry written all over her face.

Lynn listened to the sounds in the stethoscope. The raspy sounds reverberated in her earpiece, indicating significant congestion and fluid in the child's lungs. This was the twelfth child she had seen that morning. The pathogen from the high plains was moving through the local Terran tribes, affecting mostly children, though apparently a number of adults beyond those who returned on the shuttle had already caught it and six had expired from it, whatever "it" was.

"Beth, I really need the genome sequence completed on the samples I've collected. I don't know what this is, and I don't know how to treat it until I do. I don't recognize this from any of my experience with Earth viruses."

In the end, Lynn prescribed a steroid treatment, though she knew this was only going to be marginally effective. To develop an immune system booster, she needed the genetic sequence, and she wouldn't have that until tomorrow at the earliest.

The two colonies, Benston and Ammonsberg, were in quarantine. School had been canceled in both settlements as children were a vector for spreading the sickness.

* * *

Terra, who was now trapped inside Ammonsberg, was pacing back and forth at the side of Jacob's crib. Her little boy had a fever and congestion—the same symptoms many of the other children had.

Jacob had been such a hearty and healthy baby, but now he was clearly not feeling well. Ammon, who normally went out on business within the settlement, was also home and still very sick. He rested in their bedroom. He was not happy to be confined to his residence, but those were the orders of the doctors who were assigned to help the Terrans fight this illness in their community. No one knew for sure how widespread the sickness was, but the number of workers available to help with the crop duties was down to 50 percent of what it had been a week earlier. The illness had spread to the general population about three days earlier with the death of two Terran children. From that moment until now, the sequencers have been busy working on the genetic sequence of at least three main variants. The virus seemed to be one step ahead of their efforts.

Everyone in Benston assumed that once the main variants were identified, a multiple-immune-system booster would be developed and given to the population.

* * *

Dr. Harrison held the stethoscope over Jacob's chest, listening intently.

"How long has he had this fever?" he asked Terra and Ammon, who had gotten out of bed when the doctor showed up.

"Since yesterday. He hasn't slept at all, and neither have I," Terra answered.

"I'm going to give him a blood enhancer, which should lower his fever for a good twenty-four to thirty-six hours. I'm also going to give him intravenous fluids to make sure he is hydrated. I will leave this humidifier and mask. Put this on him every hour for fifteen minutes. It will help loosen up the chest congestion. From what I've seen, the illness will run its course in two to four days."

Dr. Harrison finished, then turned to Ammon and said, "I'm going to need to do a chest X-Ray on you, Ammon. I'll position this portable X-Ray over your chest. Terra and I will need to go into the other room while we run this. It won't hurt, just be still and take a deep breath in when I tell you to."

After the X-Ray, Dr. Harrison said, "You have significant congestion. Your oxygen saturation level is below ninety. I'm going to have a hovercraft come and get you. You will be hospitalized until we see improvement. You need to be intubated."

Dr. Harrison gathered his instruments up and closed his medical bag.

He spoke into his com device, "I need medical evacuation for Ammon Khan at the Khan's residence. I will accompany him, send the large hovercraft, over."

"Copy that, Doc. ETA is about ten minutes, over," came the response from the Benston medical facility switchboard.

"I do not want to leave my people," Ammon protested.

"You are coming with me; there will be no argument. Your condition is serious," Dr. Harrison firmly answered.

"What about Terra and my son? I need to be here for them!" Ammon coughed as he weakly answered and stumbled to his knees.

Dr. Harrison pulled out a small oxygen tank and mask and fixed it over Ammon's mouth and nose.

"Lie down on the floor, breathe deeply. Terra, get him a pillow and blanket. He's shivering."

"Here, my love, put this under your head," Terra said as she spread a blanket over him and lay next to him on the floor, trying to keep him warm.

"Ammon, Taylor will go and be with you. I will watch little Jacob. Sometime tomorrow, I will switch places with Taylor, and she will come here to look after Jacob, and I will come and be with you," Terra said as he was transferred to the hovercraft, which waited outside the residence.

* * *

CHAPTER 20

The Long Dark Night

"AMMON, I NEED you to acknowledge that you can hear me, I am Dr. Nick Smythe. I am your doctor."

Ammon nodded slightly as he opened his eyes.

"You are in the Benston Medical Facility. We have intubated you to help you breathe. We are also supplying a powerful medication into your lungs to help your lungs clear the congestion in them. We may have to leave this in your mouth and airway for two to four days. I know it is uncomfortable and probably frightening. We are going to give you a medication which will help you remain asleep or 'unconscious.' This will make it easier for you to handle this situation. I believe you will recover, but for now, I wanted you to know where you are and what we are doing for you. We have our best people attending to you. Rest now, Ammon. Rest."

Ammon dreamed. In his dream, he saw himself asleep in the bed they had placed him in. He saw Taylor sleeping in a reclining chair next to his bedside. He saw people, dressed in white, coming and going, doing things to him.

I can float! I am free! I am not bound to the ground!

Ammon enjoyed this newfound freedom.

I think I can go wherever I want to go. I want to see Terra and my son, Jacob. They must need me. I will go to them.

Ammon floated to the ceiling, then through it. It was night outside. There was no moon, just stars.

Perhaps one of those stars is where Terra came from? I want to go there first. I want to see Terra's home.

As he thought this, he was flying up to the sky, and then into the blackness above, only it was not totally black.

There are so many stars! They are so beautiful!

He started moving faster and faster. The stars, some of them, seemed to get closer and closer.

Soon, he came to a region of space that had fewer stars. It was darker, but there were still some stars. One star became brighter and brighter.

Perhaps this is Terra's star? When will I see her world, "Earth"? Yes, that is the name of her world, Earth. A strange name. "They"—the Earth people—call our world Terra Nova. Maybe that sounds strange to them?

He traveled onward, believing that the bright star—the one that was growing brighter—was her star.

Yes, it must be her star, her sun. The sun that shone on my beloved one!

The star became huge and bright in his vision. He imagined himself hurtling straight toward this star, her sun. Just as he thought he would plunge into this fiery orb, he flew past it and rounded the backside of it and was flung away from it. As he rounded the star, a new view was opened up to his vision. He saw a small rocky world, pocked by craters, impossibly close to this strange new sun. As he passed it, he looked forward again and saw another world. This one seemed covered by clouds; he saw no other features. Something inside of him told him this was not a good world; it was dead. He flew past this inhospitable world, and he looked forward again. Now he saw another world. This one was beautiful. It was blue and white. As he got closer, he saw features. Land, with brown and green and some white on that ground. The pieces of land took on more definition. He saw pools of blue on the land in places.

This is her world! It must be Earth! But where is her home? She said she lived in a place called "Colorado." Where is Colorado?

As he asked this question, he began descending through the clouds toward a part of the land with many mountains.

She said she lived near the mountains. Maybe this is Colorado?

He continued to descend through the clouds. He zoomed toward the mountains, then over the mountains. As he did, he saw a large city. As he got closer and closer, he became sad. This city was destroyed. Something had destroyed it.

What destroyed this city? My love told me the enemies of the Mothers had caused great destruction to come upon her world. What of Terra's family? Were they killed in this great destruction?

Ammon hovered at a great height above the destroyed city. A light below in the rubble of a building caught his attention. He flew toward the light. He floated through the ceiling and descended toward a group of people who were huddled together. It was cold. Ammon could feel the cold as if he were actually there.

"Kelly, sweetheart, do you want another blanket?" an older man asked an older woman.

"Oh, thank you, Ed. I'm freezing. My clothes are still soaked from gathering tubers outside," the older woman answered.

Ammon saw a younger woman with two young girls, probably seven and three years old. They were huddled together under a big blue tarp. The older child crawled out from underneath the tarp and walked over to the older couple.

"Grandma, you look so cold. I have some socks you can wear. They are dry," the child said.

"Thank you, Ellie, that is so kind! Grandma is freezing, especially my feet!" the older lady said to the child named Ellie.

Is this Terra's family? Ammon thought. He thought so. Somehow, he knew it was.

It was at this moment that Ammon considered what had happened to him.

I traveled through space, great distances, to Earth! How was I able to do this? Why was I able to do this?

Ammon thought, and as he did, a woman appeared to him. She had long dark hair with a jewel on her forehead.

She spoke to him, "Ammon Khan, I am Eve of Eden, the Mother of Terra's world, the Earth. I have freed you from your body so that you could see Terra's world and her family. Terra has been told her family is alive and will be brought to Terra Nova on the *Calypso.* The *Calypso* is lost at the moment in another spacetime, or universe. They will need help to return to this universe. I do not exist in the spacetime they are in. I know what needs to be done for them to return to this universe, but I cannot reach them, I do not have a starship to cross a large wrinkle. I have a physical body, and only energy—soul energy, for instance, can cross the boundaries of universes without traversing a wrinkle in the appropriate manner. Only someone who resides there and who has the knowledge to assist them can help them. I am asking if you would be willing to help them, but this means you must surrender your current body, you must die. Then you will be reborn in this other spacetime. You will grow up and acquire great knowledge as a physicist, which you will use to assist them so they can return to our spacetime. Are you willing to do this?"

"So I must die, this night, and somehow I can then assist them in returning to our universe? What will happen to my love, my Terra? What will happen to my son? My people?"

"They will be well taken care of, Ammon. Jacob will grow up to become a great leader of your people and many other peoples. Terra will grieve for you, but she will have Hatcher, who will care for her. If you choose to do this, you will be born on the other Earth. Your body will look like the body that you have now. Time and sequencing do not have the same order that you observe here. You will slowly regain your memory of your life here as you grow up. You will be given the knowledge to help the *Calypso.* Ammon, this is a great sacrifice for you to make. This act will do much to ensure that you will one day dwell with Terra on a world like mine with Terra and your descendants."

"I will do as you ask, oh, Great Mother, Mother of my beloved."

Eve smiled, and Ammon was filled with her love, so much so that he broke down and sobbed uncontrollably.

"Return to Terra Nova. I will allow you to see and communicate with Terra and your son before you pass on."

* * *

CHAPTER 21

Between the Darkness and the Dawn

AMMON WAS FLYING again, only much faster this time. He left the Earth, rising swiftly through the clouds and back to the blackness of space. The stars began to grow and then fly by as he passed them. In the distance, a star began to grow bigger and bigger until it came to a new world, his world, Terra Nova.

He descended through the night sky, toward his home where he, Terra, and his son dwelt. As he entered the room of his child, Ammon saw Terra sleeping on the rug next to Jacob's crib.

Terra stirred, believing someone had entered her home. She rose from the rug and looked around her but saw no one. In a moment, she sensed it was Ammon, but he was nowhere to be seen.

"Ammon, I sense you. Is it you that I feel?" she asked.

A quiet yet very clear voice said, "Yes."

"Why are you here? Has something happened to you? Oh, are you," she paused and began crying then whispered, "Are you … dead?"

Terra began to cry, but as she did, a warm and comforting feeling filled her. She felt Ammon's love, which comforted her.

"Why have you come here?"

Again, a very clear yet quiet voice said, "I have come to say goodbye. Eve will tell you why I have chosen to leave at this time."

Ammon looked at his son in the crib. He seemed better and was sleeping soundly. Ammon felt so much love for his baby son. He knew he would, as Eve had promised, become a great leader. His son would carry on his legacy. He was comforted for the future of Terra, his son, and his people.

He knew he must return to the hospital bed in Benston.

Ammon floated to the ceiling and flew to Benston, to the medical facility and to the bed where his body lay.

As he looked at his body, he saw several tubes inserted into his chest and noticed there was a bustle of activity around him as doctors and nurses worked frantically to resuscitate him.

Ammon entered his body, which seemed cramped and very sick. He labored mightily to breathe, but he could not sustain the effort required to breathe. He felt himself losing consciousness, he knew they had intentionally caused him to sleep, but this loss of consciousness felt different.

Ammon drifted from his body again, gazing at it one last time. He turned and looked at the clock on the wall—it was five o'clock in the morning. It was still mostly dark, but a faint glow in the east could be seen. As he looked outside the window, Ammon saw his father on his huge white horse, surrounded by other horsemen, all dressed as if they were going to war.

He reached his hand to touch the window as his father now did the same. Ammon pushed himself through the window to reach his father, who now wrapped his arms around his son.

"I have no pulse," a nurse announced to the two doctors who were exhausted, having spent the last two hours trying to save the Great Khan.

"I'm calling it; three minutes past 5:00 a.m.," one of the doctors said.

And so Ammon Khan was no more. He now belonged only to the ages. Somewhere between the darkness and the dawn, Ammon Khan rejoined his fathers and rode to the rising sun.

* * *

CHAPTER 22

Grieving, Goodbye, and a Sermon of Hope

AMMON'S BODY LAY on a bed of flowers on his ship. The children in Ammonsberg had picked the wildflowers. The crowd became silent as Terra climbed to the podium from the ground. Terra was silent for a moment, then cleared her voice.

"We mourn today the loss of our great leader, Ammon Khan, leader of the Horsemen. He was also my beloved husband and father to Jacob, who will become your leader when he reaches maturity. Our hearts are heavy, but I want to tell you what I know of life, love, and death. I believe it will comfort you.

"I know you look to me as the Oracle. I did not seek this or any other title. Even so, the Great Mother of my world, the Earth, Eve of Eden, anointed me to become The Mother of this world—our world, Terra Nova.

"We are all born with no knowledge of anything that happened before. No knowledge of why we are here, or what happens to us when our bodies die. I am here to tell you the answers to these questions.

"Our birth is actually a rebirth. Most of us have lived here before. We lived, we grew up, we found love, had a family, we worked, and then died. Sometimes we passed on because of illness, as our great Khan did. Sometimes we die in conflict. Sometimes we die in old age."

Terra paused, took a deep breath, then continued.

"Our beings are composed of two parts—a body and a soul. Eve taught me that the soul is pure energy or 'Souleng,' soul energy. This is the part of us that is actually us. All of our memories are contained in it. It can never be destroyed. It never dies. We see bodies decay after death, but our Souleng never decays and seeks an appropriate body to join itself to in our next life. There are those with greater wisdom who help us to find that appropriate new body in the circumstances where we will grow the most. This should answer most questions, but not all of them.

"Why are we here? Why do we return over and over again? We are here to gain actual, real experiences that cause us to grow and mature as souls. I could tell you a great many things about lessons I have learned in my short, young life. Perhaps you would listen and learn and avoid making mistakes I made, but probably not. It seems like there is no better teacher than actual experience, our own experience.

"We are developing a written alphabet for your language. This will help preserve the wisdom of each generation. The alphabet can be used to represent spoken words. When we string these words together, they can be written into books, which are full of words. On my world, there is a phrase called 'book learning.' This refers to things you can learn from reading books. There is another phrase called 'life learning,' this refers to learning from our experiences. The learning we obtain from our own experiences tends to be the learning we remember the best and learn from the most. Our lives offer us many opportunities to learn from either books or from our experiences, or both. We will continue to learn from each life we live, and we will become better people. We will love each other more easily. We will share more readily. We will help each other more willingly. The cycle of birth, life, and death is known as the Wheel of Life.

"Eventually, we will have grown enough that we will join The Mother's world, where we will no longer die, unless there is a special need for us to be reborn elsewhere. Apparently, this happened to me. I was born at one time on The Mother's world and then died and was reborn on Earth, and now I am here, with all of you.

"When we are on The Mother's world, we assist The Ones in developing intelligent life on new worlds. We remain there for a long time until we are ready to ascend to the world of The Ones. The Ones exist as Souleng and are able to travel anywhere in the universe. They cause worlds and stars to form and played a pivotal role in the creation of the universe with an ordered beginning or low entropy.

"Our beloved Khan was given a special mission to perform by The Mothers. He chose to leave this world so that he could perform this vital task. While we mourn his passing, we can be assured of seeing him again. This comforts me greatly, and it is my hope that you will be also. Even so, I will miss him deeply. If any of you desire more information on anything I have spoken of, please see me or my council members. I have been instructing them for some time now. Any of these women can help you with questions you may have."

Terra finished speaking and descended from the podium. A group of ten men pushed the Khan's boat into deeper water. The current from the entering stream carried the boat farther from the shore. At this point, an archer lit an arrow, pointed it upward in the direction of the boat, and loosed it, setting it on fire. As the sun set, the boat burned brightly until it began to break apart and sink. The remains of the great Khan sank to the bottom of the Terran Sea. Ammon Khan was no more.

* * *

CHAPTER 23

The Young Khan

JACOB WAS NOW twelve years old. And today was his birthday.

"You be careful with that thing!" Hatcher said, seemingly to no avail as Jacob ran outside with his new bow and arrow to catch up with William.

"I will, I will, Dad!" Jacob shouted as he ran out into the courtyard, managing to catch up to William and hop on the horse behind his older brother.

William had promised to take his younger brother hunting on his birthday. Jacob felt he had waited far too long to learn about hunting. All of his days seemed to be spent in learning various things from his mother: mathematics, physics, trigonometry, and history, which he did not understand why it was important. *What good was learning about the past when the future was where greatness would be made!* he thought.

"He'll be fine, honey," Terra said to Hatcher.

"He's just too young to be out there hunting. I *know* what can happen. Some threats you can't see coming and when you are young, you feel invincible, but you're not!" Hatcher said, with exasperation and worry dripping from his tone and words.

"He's the heir to the khan's position. He is the only child of Ammon. He is important to what we are trying to accomplish here. He will be protected," Terra explained.

* * *

"You see these smaller hoof prints? That is a deer. Wow! We need to be careful and listen and look all around us. Those huge tracks here, that is a sage cat. Those things are super quiet and absolutely lethal! I told you what happened to Dad. He lost an arm and essentially died, temporarily, from his encounter with one of these," Willie explained.

"Yes! And Dad killed it with his knife, which you have right here! We'll be OK!" Jacob said.

Willie sighed and stopped the horse. He and Jacob dismounted. Willie looked around at the tracks on the ground.

"I think the sage cat didn't get his meal. He's still hungry. Can you pull my rifle out of the saddlebag?" Willie asked.

As Jacob reached to grab the rifle, something caught his attention out of the corner of his eye.

"Will...Willie! Look out!" Jacob shouted as a blur jumped from the tree line and came running for his older brother.

Jacob tried to grab the rifle, but it was stuck on something in the bag. Instantly, he pulled his bow and an arrow from his bow bag on his back.

Willie had just seen the huge cat and backed up, tripping onto his back as he did. Jacob nocked the arrow and pulled back on the string. He knew, from talking to Hatcher, the only way to kill a sage cat was to hit it in the eye, as Hatcher had done many years earlier. This time, it was not going to be like plunging a knife into the eyeball of a stunned cat. This time, Jacob knew he had to make a one-in-a-million shot and aim for the moving cat's eyes.

The cat made one final stretch and leaped for the kill of his brother. Jacob let the arrow fly. Time seemed to slow down as the arrow flew toward the cat's head, and then it buried itself into the right eyeball of the cat, all the way to the arrow's tail feathers.

The cat fell onto Willie as he let out a huge gasp of air from the weight of the cat on his body. Jacob screamed in rage at the cat as he grabbed his knife and plunged it into the other eye of the cat. He then kept stabbing the huge animal in the chest over and over, with blood squirting from severed arteries.

Willie managed to pull himself out from underneath the cat and grabbed his younger brother's arm and said, "You can stop Jacob, he's dead!"

As he said this, his brother stopped and looked at Willie with a wild look on his face. They both collapsed to the ground, overcome by adrenaline.

"I guess," Willie paused, still out of breath, "I guess you've been practicing a lot with that thing?" Willie said as he laughed and hugged his little brother.

"Yes, but, honestly, I think I got kind of lucky," Jacob said with his normal humility.

"No, you're just a damn good shot, little brother!" Willie said as he hugged him one more time.

"What should we do with the cat's body? Can we eat it?" Jacob asked.

"I don't know if it would taste all that good, but we could take it to the guy I told you about who makes rugs out of bear hides. He could make us a rug, or maybe he could make it into a coat for you, for winter. I wonder if he could keep the head attached. That would make a pretty cool coat for a future khan!" Willie said, laughing.

They ended up skinning the cat and discarding the remains of the animal. They were both sure the hide could be made into a very warm coat. It would also make the story believable to those who could not believe a twelve-year-old boy killed a sage cat by shooting it in the eye, saving his older brother.

* * *

CHAPTER 24

Burden of Leadership

TAYLOR CLIMBED THE podium with the pleading document and handed it to Terra. Terra read through the document and then handed it to another aide, who was sitting next to her.

Terra began, "I have read the pleading. Bamli, you claim that your husband Tem left you for another woman. You claim this was because you could not give him a son. I see no pleading from him to dispute these claims. He has failed to appear now for the second time. I will rule in your favor. You are hereby granted spousal and child support for you and your daughters up to one-half of all of your joint possessions. Additionally, he is to provide you 30 percent of his yearly increases in food stores and animal stocks for the support of you and your children. I would like to emphasize a point I have tried to make several times now, and that is that males determine the gender of their offspring. It is not a 'failure' of either the male or female that determines the gender of children; there is no control over this among native Terran populations at this time."

By the time noon came, Terra was exhausted. She was pregnant again with her fourth child, her third by Hatcher.

"I am quite weary. We shall continue the hearing of pleadings tomorrow, starting at eight bells," Terra announced.

Taylor gathered up the papers left on the pedestal next to Terra and carried them with her back to Terra's residence that she had shared with the Khan at one time. She now shared it with Terra and her three children, her own three children, and Hatcher, their husband.

Taylor looked at Terra, her friend and lover, and said, "Terra, these weekly pleadings are a burden for you. Perhaps Connie could take your place. She is sixteen now, and she has your gifts. She could be a great help to you, especially where you are now with child again."

"Do you think she is ready for this? She seems like a child still. I am afraid I would be putting too much responsibility on her."

"She is ready. She wants to help you, and she is very smart and wise for her age. Let her do this for you."

PART TWO

Alternate Earths

CHAPTER 25

The Doctor is In

MARIA HAD FOUR of the most interesting strangers on Earth, *her Earth.*

To think of the idea of multiple universes was mind-boggling. "Their" Earth had "failed," and they were on their way back to rescue survivors. I want to help them. They are kind of like family. I know I should help family—that's what we do, right? They don't seem dangerous, and Carlo, well, he is kind of handsome, and he is Italian! Maria thought.

She continued to think. *How had we managed to get things "right" and why? I want to try and find out some of these things. Nothing like this has ever happened to me before! I am a pioneer, a "first" no one else has met people from another universe and another Earth!*

She brought them to her house, a mile from Idaho State University. Everyone was given a place to sleep. They all wanted to take a shower, and they were hungry. Maria tried her best to accommodate their needs.

One of them is Timeye, I think? He is an alien. He looks so human. Why is that? What happened to their world that they needed to leave it? What is their new world like? What was their spaceship like? How did it make such a long voyage? Questions! I have so many questions, Maria thought to herself as she lay in her bed, trying to go to sleep, which eventually she did.

* * *

"Levvi, how are you? Good, good. Hey, I have a special request for you. Could you meet me in my department's conference room at 1:00 p.m. today? I have some very interesting people for you to meet. They want to discuss your work on inter-universe travel. Good, I'll see you then." Maria hung up the connection on her wall computer screen.

"So we'll meet him at 1:00 p.m. on campus," Maria told everyone.

* * *

Stuart and Timeye were looking at their complimentary campus holo-pads. The writings of Dr. Levvi Markanen were expansive on the subject of many universes and how to move from one universe to another. Both were speed-reading the published works when a very tall man walked in.

Stuart stood and said, "I'm guessing you are Dr. Markanen. I am Stuart Carroll, a physicist from the FBS *Calypso*, and this is Timeye, from the planet Caldonia. It is a pleasure to meet you. We've just been reading through some of your published writings."

"Yes, Levvi Markanen am I, and am curious I about introductions of yours. Does what 'FBS' stand for and is the *Calypso* what? *Are* you, Timeye, are another planet you from? Ask must I these questions before start we. Confused feel I a bit."

Maria interrupted and said, "Levvi, speak men these old English, however, think I understand will you."

"Ah, yes. Well, FBS stands for Free Bloc Ship. The *Calypso* is an asteroid that was turned into an interstellar vessel, which we piloted to a nearby exoplanet that we named Terra Nova. Timeye is an alien from the planet Caldonia, which is many light-years from Earth. We picked him up on the journey to Terra Nova. But most importantly, we are from Earth, but not your Earth. We slipped into your universe after traversing a large wrinkle in spacetime. We need help in getting back to our universe so that we can return to our Earth and the Mars colony to rescue survivors and take them to Terra Nova. That is a lot to process, but that is our plight and status," Stuart said, smiling as he did so.

Dr. Markanen said nothing for a moment, then turned to Maria and said, "Get I it. Is first April today? Tell jokes each other and play tricks each other on this day, correct am I, Maria?"

Switching to old English, Maria said, "Well, yes, Levvi, we do tell each other jokes on this day, but I met them two days ago when I was on my way home from my mother's house in Montana. Their spaceship landed somewhere near where I picked them up. I believe their story."

"OK, Maria, for now, I will play along. So you went through a spacetime wrinkle? What is a wrinkle in spacetime?"

Timeye answered, "A wrinkle in spacetime occurred as spacetime expanded and then snapped back after cosmic inflation occurred, just after the IPT, or the Big Bang. We obtained a map of all of the wrinkles in our universe and a program that shows their evolution over time, since they are not static, though their movements take millions of Earth years. We believe many of the same wrinkles exist in your universe as the point of divergence between our universe and yours apparently occurred at approximately the year 2000 of the current era, per Earth time reckoning."

"How did you come to be in possession of this map and program?" Levvi asked.

"Life on our Earth, and we believe on yours as well, was fostered by an advanced race of aliens who now reside on a planet in the Cervantes star system. We call them "The Mothers, as it is a matriarchal society where women, usually, but not always, preside. This is quite a divergence from all of Earth's history, which was quite patriarchal. Men are the primary builders of societies and worlds, but women—certain women who possess certain abilities—preside. These women are sometimes called oracles. They can commune with the world of The Mothers even at very great stellar distances, and they can read the thoughts and intentions of people. There may be other abilities, but those are the ones we know about. The Mother of our Earth is named Eve of Eden. I have met her twice now—once on the world of The Mothers on our journey to Terra Nova and once on the Earth, though it was thousands of years before the present. The story of how we got to that time frame is also somewhat similar, I believe, in how we got to be in your universe and on your Earth," Stuart said.

Levvi shifted in his seat and cleared his throat, then said, "My mother was a linguist, like Maria. She spoke old English to me, so I shall try to speak it to you as well."

"Spacetime is one word, not two words. I believe it is because Einstein wanted to emphasize that it is one quantity, not two. It can, however, be represented as two quantities on a two-dimensional graph on a 'plane,' a sheet of paper, for example. The x-axis will be time; the y-axis will be 'space.' Of course, space is three-dimensional, not just one-dimensional like it is on our graph. When it is represented this way, it is assumed that the person looking at it knows it is really three dimensions, not just one. It helps us to convey the ideas associated with spacetime easier if we represent it that way.

"I believe that the wrinkles you speak of are like waves on an ocean, or I suppose, they could be thought of as wrinkles on a bedspread or a sheet as well. Hence, my initial lack of understanding of what you meant when you said 'wrinkles.' The wrinkles shorten the distance between two points on opposite sides of the wrinkle. This is the advantage you have been using to travel to places quicker.

"Multiple universes/spacetimes are stacked spacetimes. The distance separation between spacetimes is almost always nearly zero. The time axis is where larger differences occur between adjacent spacetimes, though it is also usually a small to moderate shift in time. The time shift between spacetimes can be ahead of or behind the spacetime you came from.

"As an aside, I want to say that time travel in any one spacetime is possible, but only back in time. Traveling forward in time is not possible as the universal waveform has not collapsed because no choices have been made that would collapse the waveform. It is possible, however, to go forward in time when traveling between universes. There are time shifts between universes.

"The reason you went back in time when you visited the prehistoric times of your Earth in your spacetime is because you traversed the wrinkle at an angle not precisely normal or perpendicular to the front side of the wrinkle. This angle is an angle on the plane or the 'page' or sheet of paper. When you hit the back side of the wrinkle, you will not be hitting it in the location and *time* you would have if the line of your travel had been perpendicular to the front side of the wrinkle surface. This caused you to travel backward in time and to a very different spatial location.

“To travel to a different spacetime, you must also change the angle again, but it is a different angle. This change in angle as you contact the wrinkle surface must be ‘upward’ or ‘downward’ relative to the plane or flat portion of the spacetime you just left. This lets you access other spacetimes above or below your universe as you travel under the wrinkle, outside of spacetime. When you hit the backside of the wrinkle, it will be the wrinkle in that new spacetime, which is time shifted and not the same shape at that time as the wrinkle in your spacetime. This means you have reentered a spacetime, but not your spacetime, nor at the time you expected.

“To calculate the angles for you to get back to your universe, we need to use a supercomputer, but first, we need to precisely determine the point of divergence.

You mentioned that you believe it was near the year 2000 CE. We will need to know this for certain. I assume you have historical records for your Earth that we can compare to historical records from my Earth and determine this point of divergence or splitting. At some point, the universal wave function collapsed to two different outcomes and then continued to evolve onward from that point. There is a supercomputer I have been using at CERN to do theoretical work on this subject. I will want to use that to determine the precise flight parameters you will need to return to your universe and your time.”

CHAPTER 26

Point of Divergence

OVER THE NEXT three weeks, Carlo and Levvi worked on determining the point in time where the history of the two Earths diverged. The resulting changes in Earth's history were stunning.

"I've found the exact wording of the US Supreme Court decision in *Gore v Bush*. On my Earth, the court decided, along partisan lines, that the Equal Protection Clause could be used to prevent counting the undervotes in the contested Florida election for president. On your Earth, Justice John Paul Stevens convinced William Rehnquist, the chief justice, that a statewide recount conducted under new unified rules would provide equal protection for each ballot, and he prevailed upon the chief justice, who then voted with the minority to give it a majority of 5-4. As a result, Al Gore won the election after winning the recount in Florida on December 29th," Carlo finished.

"So what were the 'stunning changes'?" Stuart asked.

Carlo smiled and said, "In August of the next year, CIA director George Tenet, in an Oval Office meeting with President Al Gore and his national security advisor, convinced them of the seriousness of an Al-Qaeda plot to hijack commercial airplanes and use them as weapons to destroy large buildings. President Gore issued a warning to worldwide carriers to screen Middle Eastern passengers against this threat, up to and including searches for items that could be used as weapons on board flights. He also instructed the FBI to detain the Al Qaeda suspects who were in training at US aviation schools. The result led to later convictions of the individuals and information on the mastermind of the plot, Osama Bin Laden. His capture in Afghanistan in January 2002 led to his deportation to the US to stand trial. He later died in incarceration at the Super Max Facility in Florence, Colorado, in December 2027. There was, as a result of all this, no 'War on Terror.' But the changes caused go well beyond this."

Stuart and the others paused upon hearing this, and then Stuart asked, "What other changes occurred?"

"Al Gore won reelection in 2004 and implemented sweeping banking reforms, which prevented banks from packaging mortgage bundles and selling them to investors. This prevented the financial collapse of 2008-2009. Again, something very bad was avoided—there was no 'Great Recession.' As a result, the election of 2008 favored the democratic candidate, Hillary Clinton. She easily defeated Mitt Romney. Her successes as president were transformative. The electoral college was eliminated in favor of a direct popular vote to determine the presidential winner. Once this happened, Republican candidates never won another presidential election and disappeared as a political party in 2028, replaced by the New Progressive Party headed by Alexandria Ocasio-Cortez. Again, there are even bigger changes in our history than this," Carlo smiled again.

"You're enjoying this, aren't you, Carlo?" Maria smiled as she said this.

"OK, so there were more changes. Millions did not die in the worldwide Great Pandemic of 2020. The US Senate was eliminated in the Constitutional Reform of 2030. Bribery of elected officials was also eliminated. Am I missing anything? So December 2000 was the point of divergence? What do we do next?" Stuart asked.

Levvi swiveled in his seat to face Stuart and said, "I think you and I need to visit CERN, or more specifically, my office in Geneva. My staff there can help us as we run simulations on flight parameters for your return to your universe. The calculations are prodigious. It will probably take three or four weeks. In the meantime, the rest of your crew should visit Euroflow, a maker of hydrazine and other rocket fuels. I don't know what it will cost to buy the amount you need, but you should probably work on that problem as well."

The discussion concluded on this note. Stuart and Levvi booked a rocket flight to Geneva, leaving the Salt Lake City Aerospace Port before returning to Maria's house that evening.

CHAPTER 27

CERN

THE DRIVE FROM Pocatello, Idaho, to Salt Lake City was made via hover car.

The city was significantly smaller than it had been two centuries earlier. The Mormons, most of them anyway, had left for Jackson County, Missouri, in the years from 2045 to 2070. The Mormon prophet, Joshua Oaks, announced a revelation in 2045 that Jesus would be returning to the original home of the Garden of Eden in Jackson County, Missouri, very shortly. This started the biggest migration of people in the twenty-first century. More than three million Mormons flooded into Jackson County over a twenty-year span, straining resources for the state of Missouri. By 2070, with no returning Jesus in sight and no tax-exempt protections anymore, the Mormon Church released all the General Authorities except the Quorum of the Twelve. The Twelve and their extended families continued to manage and, of course, hold on to the vast accumulated wealth. Most rank-and-file Mormons, however, were greatly embarrassed by the whole "Jesus is a no-show" and quietly moved elsewhere in the country. Very few returned to Utah, opting instead to reintegrate into American society without any religion.

The rocket flight from Salt Lake City, Utah, to Bern, Switzerland, was two and one-half hours. The drive from Bern to Meyrin, Switzerland, was almost two hours. By the time they arrived, it was almost 10:00 a.m. local time. Both Levvi and Stuart were jet-lagged to the point of exhaustion. They decided to start their work at night, using the CERN mainframes. Nighttime was "the right time" as the computers saw usage drop considerably after 10:00 p.m.

"OK, I have the starting conditions input, now I am setting the convergence criteria. Of course, this could run to completion in six hours and still not reach the convergence criteria, but that is the nature of this. It is a bit of trial and error," Levvi explained.

"You know, before we got lost, we were in daily contact with Terra Nova. The young girl, Terra Nova Winstead, who was chosen to be 'The Mother' of the new world, married another young boy on the *Calypso* named Hatcher Newsom. After we left, Hatcher and two others were held hostage by the leader of the Riders, a large group of Terrans who raped and pillaged for their maintenance. Terra was compelled into a marriage with Ammon Khan, the warlord leader of this group. He was a giant of a man, seven feet tall. We received pictures of him once on our daily com cycle. It's very strange, you look so much like him," Stuart said.

"Well, I wasn't going to say anything, but as I reached my teenage years and continuing to the present, I had dreams where I saw myself as a character such as you described," Levvi said. "I saw myself married to a beautiful young lady, but I didn't know her name. I don't place much stock in dreams and such, but I have continued to have the same dreams, nonetheless. I've even been to a therapist and discussed what this could mean. They have been of no help. Don't laugh, but I've even considered a 'hypno-regression' to another lifetime. There is a lady in Scottsdale, Arizona, who does such things. I'm going to her next month. There has to be some explanation for this."

"Maybe there is. When we visited the ancient Earth in our spacetime, we met Eve of Eden, who has a special relationship with Terra, who is now a grown woman with children. She teaches something called the 'Wheel of Lives', which instructs adherents that we live many lives, reincarnation, if you will," Stuart said.

"Do you believe in that?" Levvi asked.

"I became quite close to Eve, so yes, I believe her and I do believe in what she teaches."

"Can I see a picture of this Terra Nova Winstead?"

"Well, yes, you can. I have a picture of her, it may not be all that recent, but I have one in my personal picture file. Let me find it. OK, here it is," Stuart said, handing it to Levvi.

Levvi put the picture on holographic projection mode and waited as the picture animated itself. "I'm not sure what to say. I'm dumbfounded. This picture looks just like the woman in my dreams!"

"You're kidding, this woman, Terra, is the woman in your dreams? Terra is a highly unusual young lady. She has been called the Oracle because of her special abilities," Stuart explained.

"You said 'Oracle'? Humm, in my dreams another woman, a tall, slender, long dark haired woman with a jewel hanging from her forehead, shows up and tells me about a woman, who she calls an Oracle," Levvi said, his voice beginning to quiver.

Levvi turned away as he became emotional and began crying. "I'm sorry. For some reason, this whole conversation is resonating at some very deep level in my being. I'm sorry, please excuse me."

"It's OK. I have had to consider many changes to what I believed as a result of strange encounters on the long voyage to Terra Nova and this long voyage to rescue survivors on Mars and the Earth. I no longer feel as if everything that I can see, touch, or smell is, in fact, everything there is in the world around me," Stuart said.

"I'm going to have to process' all of this," Levvi said. "For now, I will set it aside. Surprisingly, our first attempt has just converged on a solution."

"Let's look it over together," Stuart said.

"The solution is a surprising one. It is elegant and actually quite simple. There are time parameters on this solution. It requires that the traveler pass through the first wrinkle in four-hundred and sixty-one days from now, just beyond the orbit of Jupiter. Can you even make that time frame? I mean, you need to refuel, ascend to orbit, travel to meet your starship and dock with it, all within sixty-one days. Is that doable?" Levvi asked.

"Assuming we get fuel by the end of this week and leave right away, I think so. I'll have to talk to our guidance and control engineers on the *Calypso* to be sure," Stuart said, feeling urgency in his own voice.

"Let's rerun this, make sure of the result. In the morning, we can get hold of Euroflow," Levvi said.

CHAPTER 28

Hurry Up!

"SO IT IS possible," Stuart said in his com device to Stephanie in the shuttle.

"Yes, but I'm going to have to head to Frankfurt right away to fill up at the gas station," she said, chuckling slightly at her attempt at humor.

"We'll know more when we speak with Euroflow on a conference call this afternoon," Stuart said.

At 1:00 p.m. that afternoon, they spoke with Evan Schueler from Euroflow. The fuel was going to cost 1.5 million euros, a sum they did not have.

"So what are we going to do?" Stephanie asked Stuart.

"I don't think anything we have to trade will equal or exceed that sum, but maybe I'm wrong. The plans for our miniaturized Thorium reactor might entice them. I'm going to have to see if that is something they don't have on this Earth. Can you get the schematics downloaded and sent to me?" Stuart asked.

"I have never heard of a miniaturized Thorium reactor. This powers the 'spinner', life support, and other functions on your 'shuttle'?" Levvi asked.

"Yes. The shuttle is a multifunctional landing and ascent vehicle for a wide range of planetoids and planets. The main rocket engines are used for slowing orbital velocity for descent and for ascending from a surface back to orbit. The spinners can be used for ascent and descent and transport on planets with gaseous atmospheres. Descent parachutes are available for use in slowing the descent to the surface of a planet with atmospheres. The thorium reactor is used directly or in support of these functions, especially in powering the spinner blades," Stuart explained.

Evan Schueler was kind enough to take another call from Stuart. "So this miniaturized thorium reactor performs all these functions and runs for upward of forty years on the fuel it has? I'd say yes, we are interested. I must ask, who are you that you have produced such a craft? Who funded this? And you're willing to give this away for 1.5 million euros' worth of hydrazine fuel? Why, I must ask?"

"Evan, this is Dr. Levvi Markanen. We have spoken in the past. The backstory here is pretty unbelievable and probably can't be believed until we see this shuttle craft. They are willing to fly it to your main facility for you to look at. I think that will help greatly in this instance. Then I will tell you the backstory. Sound good?"

"Ja, count me in!" Evan said.

Stephanie flew the shuttle to Frankfurt and landed in a big empty parking lot north of the main Euroflow facility. Executives and engineers were waiting to look over the craft.

"Holy Schiess! Look at the size of it!" Evan said to Stuart, who was standing next to him as the shuttle descended to the parking lot.

The shuttle passenger ramp was lowered to the ground, and Stephanie walked down the ramp and came up to Stuart and Evan.

"May we take a look at the inside?" Evan said as the engineers next to him began snapping pictures of everything on the outside of the Shuttle.

"Be my guest," she said.

As they walked through the shuttle, Stuart did his best to explain as many features as possible to the engineers.

"We could give you the entire schematics for the vehicle. There is no downside as far as I can see. But of course, I am no expert on the effects of sharing technology between different universes," Stuart said, then caught himself, realizing that he had just essentially told them who they are.

"Wait, Dr. Markanen, are these people from another 'spacetime' or universe? I know you have performed research in this area. Is this for real?" Markus Schultz, the Euroflow CEO, asked.

"Yes, that is who they are. I spent time with one of these people, Carlo, who is a historian. We determined the point of divergence was the 2000 US Presidential Election. In their timeline, the high court disallowed a full recount of a key state, giving the victory to George W. Bush. In our timeline, Albert Gore won after the high court allowed a recount of Florida undervotes. The changes from this point of divergence were stunning and dramatic, especially for the United States. In their time frame, two nuclear wars occurred, and climate change was not addressed in a meaningful way until the 2050s. Their Earth was dying by the time their starship, the FBS *Calypso*, departed the asteroid belt for the exoplanet they named Terra Nova in the Cervantes star system. Later actions by rebellious men who fought against 'The Mothers' destroyed the Earth, with but a few survivors. Colonists on a Mars colony remained alive but in a dire position. Their starship, *Calypso,* was en route to rescue these survivors when they accidentally entered our spacetime."

"I think I feel satisfied that we will certainly benefit from getting schematics on their reactor and other key systems. This is a most impressive craft. We could fuel it today. Is that acceptable?" Markus asked.

"That will work for us, Herr Schultz. Thank you. We need to leave as soon as possible. May we perform our liftoff from here?" Stuart asked.

"Um, yes, what is the safe viewing distance?" Markus asked.

"Three hundred sixty-six meters, give or take," Stuart answered.

Later, after the shuttle tanks were filled to capacity, everyone boarded the shuttle. Stuart remained on the ground, talking to Levvi, Markus, and Evan.

Levvi pulled Stuart aside and asked him in a whisper, "Stuart, I would like to go with you. Is that possible?"

"Well, you know, you may never be able to return. Are you OK with that?"

"I am. I am not married, have no children, or other attachments. What I do have is a great desire to see this woman, Terra. I feel almost certain she is the woman I see over and over again in my dreams. I need to pursue an answer to this mystery."

"All right. Get on board and get strapped in. This is going to be a rocket-powered ascent to orbit and then out of orbit so we can be captured by the *Calypso.*"

Stuart sat next to Stephanie in the copilot position. Stuart counted down, "Three, two, one, ignition, liftoff!"

The shuttle's rocket engines ignited, and the shuttle jumped off the ground and ascended through the clouds, leaving this strange and wonderful Earth behind.

PART THREE

Mars Rescue

CHAPTER 29

Becoming Acquainted with Terra

"NO, I'VE NEVER been into space before. I've always been a nervous flyer. It's the loss of control," Levvi said in response to Stephanie's question.

Levvi was seated on the flight deck because it offered the best views. This meant he was able to watch Stephanie and Stuart as they piloted the shuttle. For some people, it made them calmer to see what was happening. It made them feel like they had a small bit of control.

"Coming up on orbital insertion. We'll do three orbits then break free with a two-minute burn. From there we'll burn to escape the Earth's gravity and burn again to put us into a Hohman Transfer Orbit to go to Mars. It will be a nine-month trip to Mars, but that will put us in a position to rendezvous with the *Calypso* after both they and we come around Mars," Stephanie said, smiling at Levvi.

"I never thought I would go to space, let alone see Mars. So the *Calypso* is somewhere ahead of us.

Basically, we are trying to catch up to them after we both round Mars."

"That's right, Levvi. The *Calypso* will be trying to slow down after rounding Mars, and we will be speeding up. We will equalize our velocities, rendezvous with them, and then park our ship in the shuttle bay," Stephanie said.

* * *

"Stuart, tell me about Terra and the mission parameters," Levvi asked.

"She was ten years old when the *Calypso* left the asteroid belt. At that time, our destination was not the Cervantes Star System. Very shortly after our departure, Terra's father, and Terra, were in the communications center when a message from outside the solar system was received. Sam, Terra's father, sent Terra back to their quarters, so she never heard the message at that time. Later, Sam, our Captain, Sandra Simmons, myself, and Peter Chan, our chief engineer, opened the message and listened to it. We could only listen to the beginning portion, which was very beautiful music. It seemed like the music 'grabbed hold of us' and held us. It was almost hypnotic. We sent the message to Earth authorities to have them decode it.

"The decoded message invited us to visit the world of 'The Mothers' in the Cervantes system and to settle the neighboring world in the same star system.

Earth authorities at mission control agreed to change our destination to this star system," Stuart said as he finished up.

"How did you end up making the voyage in just twelve years, instead of one thousand plus years?"

"As we journeyed, we came to a planet orbiting a red dwarf star. The planet had lights that blinked on and off, seeming to indicate a message. We thought it might be an SOS message, so we sent a shuttle to the planet and discovered a group of aliens from the planet Caldonia. One of those aliens is Timeye, who is now married to Captain Simmons. They have a little boy who was born on our voyage back to Earth and the Mars colony. We're not sure where he was born, or when he was born, or how old he is, given all the issues we've encountered traveling back to Earth!" Stuart said with a chuckle. "Anyway, these aliens had fled their planet after a takeover by the enemies of The Mothers. They took a holographic map of the universe with showed where 'wrinkles' are at any time. Wrinkles developed as spacetime expanded after cosmic inflation. These wrinkles rebound and move through space like ripples on water. They will eventually slow down and dissipate, but they are at this time still moving through space, slowly, to our eyes and understanding because of the vastness of the universe. Knowing where these wrinkles are allows movement in space to locations on the other side of the wrinkles much quicker than traveling directly to a location without utilizing the time-saving features of the wrinkles.

"As I indicated, The Mothers have a group of men who are their enemy. This group of rebel men rebelled and left The Mothers. They call themselves the AOW or Association of Worlds. They have interfered with The Mothers efforts on our Earth, Terra Nova, The Mothers' home world, and many other places. One of their primary objectives is to capture women like Terra, who are called Oracles. Their special abilities include reading the minds of other people, putting thoughts into the minds of others, and communicating with the world of The Mothers, among other things. The purpose of an oracle is to guide civilizations in peaceful paths, in accordance with the desires of The Mothers."

"So Terra got married during the voyage to Terra Nova to a man named Hatcher. Yet she was forced to marry Ammon Khan, a Terran leader. This must have been very difficult for Hatcher, right?"

"Yes, at first, but the Riders or Horsemen came and settled next to our colony, which we named Benston. According to our updates we received, the Khan became friends with Hatcher, and he and Hatcher agreed to 'share' their marriage to Terra. As it stands now, with the death of Ammon Khan, she is once again only married to Hatcher."

"So she loved them both?" Levvi asked, clearly intrigued by the marital history of Terra.

"As I understand it, yes. She mourned the death of Ammon greatly, I am told."

"Well, I have bored you with my questions. I feel that I am, or was, Ammon Khan, which explains my fascination with Terra. It will be many years before I meet her. I hope to speak to her once we get back to your universe. She is so interesting and so beautiful. I must admit I am smitten by a woman I have never met."

"I can understand that. She truly is a remarkable woman, and many men have been thusly smitten as you have, though your connection to her is unlike anything I have ever encountered."

CHAPTER 30

The Space Ballet Near Mars

"CALYPSO, SAY AGAIN," Timeye said.

"We have you on long-range radar showing you are behind us by 300,000 kilometers. Your velocity exceeds ours by 20 percent. Ours is decreasing by 1 percent every hour. Intercept in sixteen hours. We will manually match velocities until the final hour. Then, the AI on the *Calypso* will take over and will manage the equalizing of both our velocities and perform docking during the final hour," Anna Stetson said.

"Copy that, *Calypso*."

"We wait now," Timeye said.

For the next six hours, Timeye and Stuart sat hunched over screens on the flight deck of the shuttle. Strictly speaking, this was not necessary, but it kept both of them occupied; otherwise, the waiting would quickly turn boring, and boredom was not a good thing during a complicated maneuver in space.

Levvi sat in the third pilot seat, which was brought up from below so he could watch and continue practicing with the flight simulator. He had become proficient in the systems of the shuttle during the previous nine months enroute to the *Calypso*. His seat and controls were only linked to the flight simulator, so there was no danger to the delicate operation that Stuart and Timeye were currently engaged in.

"What are you practicing, Levvi?" Stuart asked.

"I'm practicing what you are attempting, only I am trying to do it manually without the AI performing the final steps to docking in the shuttle bay of the *Calypso*."

"How's it going? Crashed the shuttle yet?" Timeye asked.

"Just the first time. Every other time I got us into the shuttle bay in good shape!" Levvi answered, smiling broadly, clearly proud of himself.

"Well, it is not a maneuver for the faint of heart. Personally, I don't trust the AI as much as I should. I know that is not logical, but it is borne of my unfamiliarity with this AI," Timeye said.

“I’ve got a fault light on the yaw rockets,” Stuart said. “Probably a faulty fuel tank sensor. I’m going to run a diagnostic on the fuel tank sensor system.”

A moment later Timeye asked, “What is the prognosis, Stuart?”

“Faulty sensor, which is not good. We anticipate using 85 percent of the fuel in the yaw rocket tanks,” Stuart answered.

All three men were silent for a moment, contemplating what could be done.

“We’ve got only one chance to fix this before we get to the final hour. The primary fuel tank sensor is accessible only from outside the shuttle. I know we have a spare sensor below. Replacing the sensor is a two-man job, especially considering we are in space and our velocity is changing; we are accelerating. We’ll have to do a completely “rigged” spacewalk. I know I can do this. I am certified and you are also, Timeye, but we need at least one qualified pilot in the cockpit, just in case, you know, anything happens to one or both of us. We can’t call on the other two qualified pilots, Stephanie and Cassia. They are both in labor at this moment,” Stuart said.

“I can sit in the pilot seat. I feel confident in what needs to be done, if it came to that,” Levvi said.

“Uh, well, this may be our only choice. We need to know precisely what our fuel level is for this maneuver,”

Stuart said. "OK, Timeye and I will go suit up. You sit in my seat. Be ready."

* * *

A moment later Stuart apprised the *Calypso* about the sensor problem. "*Calypso*, we have a problem. We have a faulty yaw propellant tank sensor. We anticipate using almost all the attitude control rocket fuel. We will be cutting it close. I don't want the AI getting faulty fuel tank readings and aborting the docking. Even if we did this manually, we need to be sure of our fuel status. Timeye and I are going outside to replace it. Not a simple replacement given our acceleration status. Stephanie and Cassia are both in labor, believe it or not, at this moment. We are out of qualified pilots, and we only have two of us who are certified for spacewalk maintenance. Dr. Markanen has been training extensively on the flight simulator AI, including performing the manual docking maneuver we are going to perform in a few hours from now. Our only choice is to put him in the pilot seat. Timeye and I will do the spacewalk and sensor replacement. Over."

"I don't like this, Stu," Captain Simmons responded over the ship-to-ship coms. "You two be careful."

"Copy that," Stuart answered back.

* * *

“Goddamn! Nothing ever ends up being easy,” Captain Simmons said to Anna Stetson.

“I know you are worried, Captain. But they both know what they are doing. They should be fine,” Anna replied, trying to help her captain feel better.

* * *

“Hooking up. I’ve got the line motor on, to help pull us up to where the yaw tank access panel is,” Stuart said.

“I’m hooked up now also,” Timeye responded.

“I’m at the access panel now. I’ve got the drill out and am removing the fasteners. So far so…,” Stuart stopped midsentence as sparks from the drill erupted and burnt the drill out. At the same time, the outside lighting on the shuttle went out on their side of the shuttle.

“Well, that’s not supposed to happen. Oh, great, my suit-grounding strap is, well, not there! Why?” Stuart asked, clearly confused at the events.

“It doesn’t matter now, Stu,” Timeye said. “We’ll have to proceed as is, no drill. We can muscle the fasteners off with a screwdriver.”

“Proceeding. A gorilla must have torqued this last fastener! Dammit! I poked my suit! Losing suit pre …,” Stuart stopped midsentence and passed out as his suit pressure dropped below five psig.

“Levvi, we have a problem. Stuart’s suit has been compromised. He’s unconscious. The logical thing to do is to finish this replacement. I’m applying suit cement to his finger and glove to stop further pressure loss," Timeye said.

A few minutes later Timeye said, "OK, I've got the last fastener off, and I've pulled the old sensor out and put the new one in. Electrical checks are good." Timeye's comments were heard by the shuttle and the *Calypso*.

"Returning now. Wait! The line motor is not working. I'm going to head back. It's going to be hard, working against the shuttle's acceleration. I'll use the hand winch once I get to the airlock to pull Stuart back in," Timeye said, breathing heavily and sweating from the exertion against the shuttle's acceleration.

"That seems impossible," Levvi said. "He weighs ninety-five-plus kilograms, and we are pulling almost two g's now trying to catch the *Calypso*."

"Caldonians can intentionally draw upon releases of adrenaline in fight-or-flight situations. I can do it," Timeye explained.

Back inside the airlock, Timeye worked the hand winch furiously, trying to get Stuart back inside so he could be treated. As he pulled Stuart into the airlock and closed the door, he pressed the re-pressurization button. Suddenly, he collapsed next to the unconscious Stuart. On the other side of the airlock door, two crew members pulled both men inside and worked quickly to remove the helmets so that both men could breathe. Steve Kinney, the shuttle paramedic, put oxygen masks on both men.

"Timeye is breathing, heart rhythm is good. Stuart, not good. No pulse. Starting CPR. Get me the defibrillator!" Steve shouted.

A crewman pulled the defibrillator off the wall mounts, and both of them worked to pull off Stuart's spacesuit so they could put the defib paddles on his chest.

"Stand clear, shocking," Steve shouted as adrenaline raced through him. "Shocked, no pulse. Let's try again. Up the voltage!"

A second shock was sent through Stuart's lifeless body, with the same result.

"Continuing manual compressions!" Steve shouted.

A moment later, Steve, now exhausted, said, "Let's try one more shock, maximum voltage!"

"Clear, shocking!" Stuart's body jumped again. "Checking, I have a weak pulse. Help me get him to the med bay," Steve said as they transferred Stuart onto the stretcher and moved him into the med bay.

"He has a rhythm, but I don't know about his neurological status. We'll have to wait until we get to the *Calypso* to assess that," Steve said over the com to Levvi on the flight deck and to the listening *Calypso* crew members.

"What is Timeye's status?" Levvi asked.

"He's in no shape to help you if that is what you are asking. He's disoriented and weak, though he is stable."

"Copy that. *Calypso,* this is Levvi. I assume you heard we are down to no certified pilots. It's just me. I'm proceeding by continuing to match our velocity. Can't wait for the AI to take over the final steps at the one-hour mark. Over."

"Copy that, Dr. Markanen. Good luck, let us know what we can do to assist you," Anna Stetson responded.

By the one-hour mark, Levvi had nearly equalized the velocity with the *Calypso*. When the one-hour mark finally came, Levvi said, "*Calypso*, I am turning the shuttle over to your control, over."

"Copy that, Dr. Markanen. Activating the AI docking procedure," Anna replied.

A moment went by then a concerned Levvi asked, "*Calypso*, why is nothing happening? Over."

"Unknown, Levvi. Could the electrical shorting Stu and Timeye experienced have damaged the com lines to your systems?"

"I don't know, but I am operating under the assumption that it has," Levvi answered. "Proceeding with manual docking."

For thirty minutes—thirty excruciating minutes—Levvi coaxed the shuttle into position outside the giant shuttle bay doors on the *Calypso*. As the shuttle came into the bay and set down, he powered the shuttle down and waited for the all clear on the shuttle bay re-pressurization.

"All clear, Dr. Markanen. Lower the ramp. We need to get to your injured crewmates," Anna said.

CHAPTER 31

Here Goes Something, We Hope!

"STUART, CAN YOU hear me? We have given you a stimulant to wake you up. I am Dr. Andre Le Cesne. I am taking care of you. You are on board the *Calypso*. You lost pressure in your spacesuit while replacing a faulty fuel sensor on your shuttle."

Captain Simmons, Timeye, and Carl Carroll, Stuart's brother, waited to see if Stuart would respond.

Dr. Le Cesne scratched his beard and said, "We can't deduce anything from his nonresponse. It can take days, even weeks, for a person in his condition to regain consciousness. Meanwhile, the nanobots we introduced into his brain are delivering stem cells to the locations where we see some damage. They are also repairing capillaries that were damaged as the heart tried to deliver blood to the brain. As I said, this effort by the heart to pump blood at higher and higher pressures frequently leads to hemorrhaging. In Stuart's case, the damage appears to be minimal. I am still optimistic he will regain consciousness and use of his faculties."

Stuart dreamed. He dreamed of Eve and their night together. It was a good dream, the best dream. He had made love to the woman of his dreams, a woman who was everything to him. If he ever saw her again, he would make sure he never left her side again. She was somehow familiar to him in a way that no other woman was. He just wasn't sure why this was the case.

"Stuart, my love, you do not remember, do you?" Eve asked.

"I can't remember, but it feels so close."

"You and I lived together many centuries ago. You were my first husband. You were called Adorn. Our descendants confused your name and called you Adam. I was known by my name, Eve. Are you remembering things now, Stuart?"

"Yes! I remember your parents. I remember Eden. It was so beautiful. We had children. Seff, I remember him. He looked like me!" Stuart said as he began to shed tears.

Stuart's tears were noticed by Dr. Le Cesne, who had stayed behind with Carl after the captain and Timeye had left his bedside.

"Stuart, can you hear me? Are you dreaming? Stuart, wake yourself!" Dr. Le Cesne commanded.

Stuart pulled himself out of his dream, which was so difficult as he did not want to leave it. He exhaled deeply, on his own, through the breathing tube. He reached for the tube that was in his throat and was going to pull it out, but Dr. Le Cesne held his hand and pulled the tube out.

Stuart gasped and tried to speak, but his words were very soft. He breathed two deep breaths and said, "I am back. I dreamed of Eve. I was her husband in Eden. I want to see her again!"

Carl and Dr. Le Cesne looked at each other and smiled and hugged each other.

"Stu, it's Carl. We are so happy you are back! How do you feel?"

"I'm not sure how I feel. So we got the shuttle back to the *Calypso*? Timeye was able to do it?"

Carl answered, "Actually, Stu, Dr. Markanen was able to do it by himself. Timeye collapsed from the exertion of bringing you back inside. The com links to our computers were damaged, and Levvi had to manually bring the shuttle in. It was a good thing he had practiced doing it manually on the simulator!"

"Well, I'll be damned! Levvi is a fast learner. I guess we can use him when we are bringing people up from Mars and the Earth to the *Calypso*."

"Stuart, you should rest for a couple of hours. Later, when you wake up, I want to do an evaluation on you to see how well your mind and body function," Dr. Le Cesne said.

* * *

Stuart continued to recover over the next few months. He had to regain his ability to walk and perform other motor functions. By the time the big wrinkle transit came he was nearly recovered, though walking and moving around would tire him out easily and he sometimes used a wheelchair when he was exhausted.

"Good to see you up and about, Stuart," Captain Simmons said as Stuart was wheeled into the Guidance and Propulsion Center. "I hear you can stand on your own now. Won't be long before you are dancing again, I suppose?"

"Well, that will be a while, but I am hopeful," Stuart answered. "So I guess we are nearing the big moment when we pass through the Jovian super wrinkle?"

"Yes, about thirty minutes from now. We've got the course entered into the G&C computer. I think, and hope, that all we need to do is watch and wait, and hopefully we'll get back to our little universe soon," the captain said.

Thirty minutes later, the *Calypso* entered the wrinkle.

"Lots of vibration and rumbling, like last time, I'd say," Stephanie noted.

"Let's hope that is a good sign," Captain Simmons replied.

"Three minutes into the wrinkle, two more minutes and twenty seconds to go," Anna Stetson announced.

As the time passed, the rumbling subsided and the ship emerged into a view of Jupiter, which was now much smaller as it was much farther away.

"I see Neptune, Uranus, and Saturn," Levvi said.

"Well, whatever we did, we are in our solar system. Question is, is it our time and our unique solar system?"

Captain Simmons asked rhetorically. "Let's ping the Europa data link."

"Pinging, ma'am," Anna Stetson said. "Should take about twelve minutes round trip to get a ping back, at this distance."

"While we wait, are we able to pick up any radio transmissions from Mars or the Earth on established frequencies?" Captain Simmons asked.

"Let me check," Stephanie said. "I think I might have something. It originated on Earth and seems to have been directed at the Mars colony. Here is what I have."

"Water reclamation is down; we're back to boiling water. Our transmitter is working intermittently. Food stores are depleted. We've picked everything in a fifty-mile radius. Over."

"That could very well be survivors on Earth letting Mars know about their status. At least it seems like that to me," Stephanie said.

"I concur," the captain replied.

"OK, ma'am, I have a return on the ping to the Europa data link. It shows an Earth date and time of 17 December 2366, time 1100 hours GMT. That is only five years from when we left Terra Nova! How could that be? We've been gone seven years!" Anna exclaimed.

"Relativistic effects. Recall we did travel much faster than 4 percent of light speed, transiting two of the wrinkles. Additionally, perhaps we have moved back in time some more by transiting this last big wrinkle?" Timeye speculated.

“I think Timeye is correct,” Stuart said. “Let’s try and contact Terra Nova. See what date and time they have. Undoubtedly, they have been wondering where we are.”

“What is our estimated arrival date at the Mars Colony, Ms. Stetson? Also, I do not want to contact the Mars Colony. We don’t want to tip the AOW off to our position and time of arrival. Eve warned us of two individuals whom the AOW is compelling to spy on our arrival time so that they can destroy our vessel. I’m wondering if we want to send a shuttle to Mars and land unannounced. Perhaps this can help us get to those two people before they alert the AOW to our arrival. We need to think this through and plan it out. Stuart, can you look into how we would do this?”

“Of course,” Stuart answered.

CHAPTER 32

Stealthy Approach

"WE WILL LAND on Mars with a similar approach to how we landed on Levvi's Earth. The *Calypso* will be in an elliptical orbit around the sun and Jupiter. When we pass near Mars, the shuttle will depart the *Calypso* and land on Mars. We want to land at night and out of visual range of the colony. We don't want anyone to see us. We received information from Terra Nova about a ship that was sent to follow an AOW ship that left the moon just before the Earth was destroyed by impacting lunar crust. That ship had three occupants: Tariq Shepherd, Shandra O'Loughlin, and Carl Peterson, a trillionaire from Earth who was indicted in the attempt to destroy the *Calypso*. This individual was killed on the voyage, allegedly in a suicide attempt.

He and a small group of very wealthy individuals had made contact with the AOW, who instructed them to destroy the *Calypso*. That attempt was made by Alex Cornelius, which I am sure you remember. I believe the two returning members of that ship were captured by the AOW, and they are the two persons who were coerced into helping the AOW by alerting them on the return trip of the *Calypso*. We must not allow either of those two to know about our arrival in the shuttle. Once they are detained, we can then direct the *Calypso* to enter Mars orbit and begin the evacuation," Stuart explained.

Captain Simmons listened intently, then said, "All right, that sounds like a good plan. Stephanie and Cassia, you two will pilot the shuttle. I will go with you as well as Dieter. Our little stealthy voyage will take us three weeks. While I am gone, Timeye will be the acting captain. When do we leave?"

"Three days from now," Stuart answered.

* * *

Twenty days had passed since the shuttle Armstrong left the *Calypso*.

"How do we evade the Martian radar?" Captain Simmons asked Cassia.

"We are not sure if it is even working. The equipment may have malfunctioned. We know that a great many systems are simply no longer working as they have had no resupply from Earth for years. Our steep descent would look like a meteor if it were picked up by functioning radar. They still may have optical means—telescopes—at their disposal. They would have to be looking at the right moment in the right location to see us. Our descent begins in daylight and ends after darkness. They are not likely to see us visually as we begin a daytime reentry, and by the time it is dark, we will be on the ground. It seems to me that there is very little risk of us being spotted. We will approach in the crawler, in darkness, and literally knock on one of their airlocks. We won't use radios. The two surviving crew members of the probe that followed the AOW ship were communications specialists. They might be at a console even at that time of night. Once we get inside, it becomes a little trickier. We need to get out of our spacesuits as that has *Calypso* emblazoned on them. Our clothing must not identify us in any way, so dress casually. Everyone on that station is likely wearing clothing that is wearing out by now."

* * *

"Initiating de-orbit burn," Cassia announced.

The shuttle rumbled from the rocket engine burn meant to slow them down and allow them to begin a steep descent. The burn was longer than normal so that their orbital speed would decrease dramatically, allowing for a steep drop to the surface.

"Deploying thin atmosphere parachute. This will take us to three thousand meters, then we will burn through to touchdown," Stephanie spoke through the com microphone to Captain Simmons and Dieter.

Captain Simmons looked out the viewing window from the passenger seats she and Dieter were sitting in on the flight deck. The sky was black now, night had fallen.

"I hope they don't see our rocket burn," Dieter said. "This low-light condition seems like it would be perfect for them to see us."

"Not likely, we have a pretty tall mountain between us and the station," Cassia answered back.

"So we'll have to travel around that large mountain in the crawler to get to the station?" Captain Simmons asked.

"No, ma'am. We will get to several hundred meters and traverse laterally around the mountain at low altitude, then set down. It's the best we can do. I think no one will be looking for us at this time of night. There is simply no reason to be looking for us. They aren't expecting us," Stephanie explained.

Several moments later, the shuttle landed, still partially shielded by the hills.

"We need to get moving," Stephanie said. "We can't waste time sightseeing. We have to get the crawler out and get ourselves loaded into it. Cassia is going to remain behind with the shuttle. If we need her to fly it closer to the station, once it is safe to do so, she can do that," Stephanie said.

Thirty minutes later, they were loaded into the crawler and began the one-hour drive to the station. As they approached the station, they were surprised at the lack of lights anywhere in the station or around it.

Stephanie was driving, and as they came to within two hundred meters of the station, she commented, "There just aren't that many lights on. I wonder why."

"They are probably conserving energy and everything else. I'm beginning to think they probably don't have a functioning radar or other non-essential systems working," Captain Simmons noted.

"I see an airlock entrance almost directly in front of us," Stephanie said. "I hope there are people nearby who might hear our knocking."

All three walked to the airlock. Stephanie tapped a number of times on the door with a hammer they had brought with them. While she did, Dieter put his hand on the airlock door and could feel the vibrations from her tapping.

"Did you just tap 'shave and a haircut, two bits'?" Dieter asked.

"Yes, I did. I wanted the knocking to sound like something manmade. Just knocking might not be noticed as manmade," she said.

"Good thinking, sweetie," Dieter said as he laughed.

A couple of moments went by, and then a flashlight shone from the inside through the airlock viewing window. Stephanie turned on her flashlight and waved back at the person inside. Another moment passed, and then a voice was heard through their suit com set.

“Identify yourself. We were not expecting to have anyone outside this time of night,” came the reply.

The captain, Stephanie, and Dieter looked at each other, wondering what to say. They did not want to explicitly identify themselves yet, but they had no choice.

“I am Captain Sandra Simmons, commanding officer of the FBS spaceship *Calypso*. We're here to rescue you. Please do not announce our presence yet until we are inside and can explain why we don't want you to do this yet.”

CHAPTER 33

A Careful Introduction

"OK, NOW PLEASE tell us who you are," the man inside asked.

The captain answered. "We are trying to avoid a couple of people who work in your facility. Those two people have been compromised and would present a danger to our ship if they found out we are here."

"I've forgotten my manners. I'm Jack Wolfe, station security officer."

"Jack, the two people who we do not want to see us are Tariq Shepherd and Shandra O'Loughlin. They were on a ship, the...," Jack interrupted Captain Simmons.

"Yeah, I know who they are."

"We have reason to believe they were detained by the AOW and forced to cooperate with them in an attempt to destroy my ship when we arrived to rescue the people here at this station," the captain explained. "We can't let them know we are here; the AOW may have ships nearby that could destroy our ship. I suspect they have a miniaturized EPC device implanted into their heads. They may be acting against their will. They may not want to cooperate but are being forced to do so."

"I see. Well, let's get your suits off. It has your ship's name all over it."

After removing their spacesuits, Stephanie asked, "Will our clothes 'fit in' with the other people around us?"

"Yes, but you don't smell bad, that could give you away!" Jack smiled as he said this.

"Yes, I wasn't going to say anything, but it smells funny in here," Dieter said.

"Piss—everybody and everything smells like urine. We drink purified urine, so when we sweat, and we do that a lot, the whole station smells like a urinal," Jack said, smiling again.

"Well, we won't hold that against you!" Captain Simmons said, smiling and laughing.

"Let's get you to Kwame Olajuwon's quarters. I know he is the person you communicated with before you all dropped off the radar, so to speak. He'll be surprised to see you. Well, everyone will be. We thought you weren't going to make it. We figured we would all have to walk out of the airlock without suits at some point in the future. It really is good to see you all! Thank God!"

They were led to Kwame's quarters. They ran the ringer on his door. A moment later, a muffled voice said, "Just a minute. Let me put on a robe."

The door opened, and Kwame stared at the three visitors.

"You're Captain Sandra Simmons! I don't know these two, but how is this even possible? You stopped communicating with us a while back. We assumed you had all died in some accident or whatever. Anyway, even if you showed up, I never expected to see you this soon. We thought it would be a few more years before you arrived. I'm so very, very happy to see you! Come in. Let me call the station director, Gus Johnson."

"Wait, before you do that, we need to talk to you, we need to let you know about an insider threat that you have," Captain Simmons said.

"What are you talking about?" Kwame stammered.

"We know you sent a manned ship to follow a possible AOW ship. We have been told by Eve of Eden that two individuals were forced against their will to notify the AOW as our ship approached Mars so that they could destroy it. Now we are here, years ahead of schedule, so that helps, but we don't have all the information we need to know if our ship is completely safe from AOW threats. We need you to apprehend these two individuals. They may have implanted EPC devices that allow them to communicate with the AOW. I'm not sure how to do this, perhaps they will need to be sedated so that their inadvertent thoughts don't get communicated to the AOW," Captain Simmons said.

"I see. Can I notify the station manager now? He'll need to know about this so these two can be safely apprehended and sedated. We need to try to save them. I know the two you are talking about, Tariq and Shandra."

CHAPTER 34

Nullifying the Threat

"TARIQ," KWAME SAID, "can you come into the observatory conference room?"

"Sure, just a second."

Kwame held the door open for Tariq, who walked into the room and was immediately jabbed in the neck and given a powerful sedative. Kwame caught him before he hit the ground.

"OK, I want this room locked. Dr. Dillworth will stay here with him," Gus said before he left the room with Kwame.

"Same approach, we'll ask Shandra to come into the communications conference room. That's across the hall from here," Gus explained.

* * *

"Shandra, can you come into the conference room?" Gus asked.

"OK."

As she came through the door, she was jabbed in the neck with the same sedative.

"All right, let's get both of them moved to the medical wing. The medical staff is going to take X-Rays and MRIs of their heads. We think that is the most likely location of any device put into their bodies," Gus explained.

* * *

"I can already see the device," Dr. Dillworth said. "Just under the skin on the left temple. I must ask this question: Won't the AOW know that they are sedated and later, if we remove these devices, won't they know that also? Could that alert them to the idea that something is up?"

"Yes, that is possible. That is what Maya is here for. She is a neuroscientist. She thinks we can transmit EEG signals from their heads into the device, which will trick the device into thinking everything is still working," Kwame explained.

"So there are leads leading to at least seven locations in their brains. Some are pain centers, the others, well, I am not sure of. The only way to remove these leads is to have nanobots pull them out, so we can get them. If I were to pull the leads, I would cause significant brain trauma, bleeding, possibly even a stroke. This is going to take up to twelve hours to do this," Dr. Dillworth said.

The next day, Captain Simmons, Gus, Kwame, Maya, and Dr. Dillworth met in the communications conference room.

"It worked. The bots got the leads out, I took out the device, and then Shandra coded out and died," Dr. Dillworth said, as he broke down. "I need a minute. Tariq is doing OK. I don't know what happened with Shandra. The pain center readings went off scale, and she went into cardiac arrest. Tariq is taking it very hard. He and Shandra were a couple. He's able to talk freely now, but we have to turn off the leads and replay EEG readings from a previous hour so as not to make anything look suspicious. He told us everything. Carl Peterson was killed by the AOW; they told Shandra and Tariq that they don't tolerate failure. Apparently, destroying your ship was Carl's mission, and he failed. Ruthless, nasty bastards!"

"So what do we think they will think when they realize they aren't getting her thoughts anymore?" Captain Simmons asked.

Kwame answered, "They might think she was going to reveal her secret, and the device stopped her. They believe they still have one asset, Tariq. That is probably good enough for them, and they'll figure Tariq got the message that they don't mess around," Kwame speculated. "Still, it just hurts that we couldn't save Shandra. She was such a sweet young woman. Damn them!"

CHAPTER 35

Evacuation

GUS STOOD UP and began speaking to the assembled personnel.

"Everyone, your attention please! Captain Simmons has informed me they have three shuttles for the evacuation. Their ship will establish Mars orbit in twelve days from now. Between now and then we need to get everyone and everything ready. There are a little over five thousand of us to be evacuated. The shuttles will need to make about sixteen trips from the surface up to the ship to evacuate everyone. There are medical checks that will need to be completed on all of us prior to going into hibernation. Those with children under eighteen years of age will not be allowed into hibernation until their children reach that age. Fortunately, there are only thirty-one individuals with children under eighteen. I'm told the trip duration to Terra Nova is about nine years' time. I am not sure that it has been determined if it will be possible for any of us to cycle into and out of hibernation. That will be worked out while we are enroute to Earth, which is about a seventy-one-day trip. The belongings you can take are limited to no more than twenty-four kilograms and must fit into approximately these duffel bags that I am holding up. Our doctors will begin medical checks starting tomorrow. You can sign up for time slots by accessing the Mars shared drive under the evacuation guidelines. Any questions?"

Twelve days later, the first of the three shuttles landed on the landing pad north of the colony. Stephanie, Cassia, and Levvi would be the primary pilots for the shuttles. Cassia piloted the first shuttle.

The first group of three hundred colonists was jammed into the boarding area, which was never intended to accommodate that many people.

The PA announcer said, "Your attention: We ask for your patience and understanding. Form a single file line based on your randomly assigned number, which you were given. Proceed orderly when directed to do so. Parents with minor children may proceed with their children."

Benjamin Olajuwan was antsy. "Mommy, I'm so excited for our big trip! I heard they have some cats on the big ship. I haven't seen any cats here. I wonder if I can have one to take care of in our apartment. I miss Fritz. I don't know what happened to him. Mommy, I hope I can meet Terra, she sounds so cool. Daddy told me about her. She can read minds!"

"I miss him too, Benji. I miss a lot of things. I'm sure you will be able to meet Terra. I hear she is a very nice lady. You know, you've been such a big boy about all the things that have happened to us. I love you, you know that, right?"

"I love you too, Mommy. I'm so lucky to have you and Daddy. We wouldn't be alive if we had not come to Mars for our vacation. I don't want you or Daddy to die, ever!"

"Well, everyone dies eventually, but it will be a long time from now. Let's put your toys back in the bag, we need to go and line up. We'll see Daddy in a few days. He has some work to do here still."

A few minutes later, they were walking down the hallway leading to the shuttle. A female crew member helped direct them to their seats on the lower level of the shuttle. She showed them how to put the five-point harness on and asked them if they needed anything.

An hour later, everyone was on board, and Cassia announced their departure. "Good morning. My name is Cassia Kim. I will be piloting the shuttle today up to the starship *Calypso*. Our trip will take about two hours before we are inside the starship. You will need to remain seated and buckled up at all times. If you really need to use the restroom, you can, but we need you to raise your hand and wait to be helped to the restroom by one of our crew members. We will be experiencing zero gravity for part of our trip up today. Before you stand up to go to the restroom, a crew member will help you by giving you magnetic boots so that you can move around more easily. Stay on the marked pathway so that your mag boots can be effective. But it really is best if you try to stay in your seats for the duration of the flight today. I hope you all enjoy the flight."

The AI voice announced the countdown, "Three, two, one, liftoff!" The shuttle rumbled, and everyone was pushed down into their seats as the shuttle engines began accelerating the spacecraft.

"Mommy, will we ever come back to Mars?" Benji asked his mother, Alice.

"We are going far, far, away from the Earth and Mars, Benji. We won't be coming back. Our new home will be Terra Nova, a brand-new world."

Benji sat in silence. His mother looked at him out of the corner of her eyes. She noticed he was crying softly. She reached over and put her arm around him and pulled him closer, kissing him on top of his head.

When they finally entered and landed in the hangar bay, everyone was excited to see the ship that they had all heard about, but now they were seeing it with their own eyes. None of them ever dreamed they would see this marvel of engineering.

"Everyone, I am Crystal Seamons. I have pairs of magnetic boots for you. Please line up over here along the white line on the floor. A rope has been set up to keep all of you somewhat tethered to the floor. Put these shoes on and proceed to the entry door by following the white arrows. In the hallway after you leave the hangar are crewmembers who will take you to your quarters, but only if you have children under eighteen. Everyone else, please follow the crew members who will guide you to the hibernation chamber. Once you reach the hibernation chamber, medical personnel will assist you with the preparations for entering hibernation. Yes, you guessed it, that means a bowel cleanse! Don't worry, I survived it and so will you! Once you have completed the medical procedures, a crew member will be able to show you around the ship a little bit. I wish you could have more time to see things, but the procedures call for you to enter hibernation three hours after completion of the medical procedures. The short tour is to help keep you occupied while the cocktail of medicines begins to be absorbed by your body. By about three hours, you will be sleepy, and at that time, we want to get you into the hibernation tubes."

CHAPTER 36

Someone Else Is Coming to Mars

THE EVACUATION TOOK sixteen days. On the last day, Cassia was again on the surface, picking up the higher-ups in the station management structure, along with a number of technical people who were part of an elaborate deception plan that was being put in place to deceive the AOW.

"*Calypso*, I have everyone on board. Request permission to return, over," Cassia said.

"Negative, Cassia. We've picked up something from Europa radars operating on a dead-man switch. We will be going dark until we determine who this might be. Request you go dark and initiate operation *Traffic* immediately. Advise your passengers to return to the station and turn the lights back on, over," *Calypso* coms responded.

"Copy that, *Calypso*."

"Have you ever had any communication from the AOW, Kwame?" Cassia asked.

"Never. We have assumed they weren't in the neighborhood. Most of our comms were with your vessel and survivors on Earth, never with anyone else. I'll advise everyone of the change in plans. We'll get our visual detection systems up and running again. I assume we don't want to pulse space with long-range radars?"

"Negative, let's make it seem like we are unaware of anything or anybody," she answered.

Kwame went to the lower deck to announce the change of plans to everyone.

"Everyone, there has been a slight change of plans. The *Calypso* has picked up something on Europa radar, operating on a dead-man switch cycle. We are assuming it is the AOW. Because of that, we are going to delay our return to the *Calypso*. We're going to initiate operation *Traffic*, and we're going to make the station look alive. We want lights on, etc. I can't say what this threat will be like exactly. We've never, yet anyway, picked up something on any radar. Anything is possible. Maybe they are coming to look at the real estate here on Mars. We have to be ready for anything. I want Buck Stilson to be in charge of our defensive posture. If they do land, we might need to defend ourselves."

* * *

"Buck told me you have something on infrared. What does it look like, Ginger?" Kwame asked.

"It's a ship. It has an ion trail, and the path it is taking will bring it here in about twenty days. It is not sending out any communication as far as we can tell," she replied.

"If we somehow determine it is hostile, is there anything we can do to disable it?

"The shuttle, or shuttles, could land and hide on Phobos, the larger of the two Martian moons, and wait until it comes close enough to use rail guns. But I believe that may give away that we are up to something. They may get a message out that the *Calypso* is nearby, and that could scuttle any rescue on Earth. If they enter orbit, our best chance may be to capture them here, at the station. That way, we can learn about their plans and force them to send messages that dissuade their superiors from sending reinforcements," she said.

Kwame thought for a moment. "I agree. But just in case, let's plan on putting a shuttle or two on Phobos as a last resort."

CHAPTER 37

Big Surprise!

EIGHTEEN DAYS LATER, the mysterious vessel was still approaching Mars. It had not changed course. This is where it intended to go, that much was clear. The Mars colony had played dumb and had not pulsed nearby space and had sent no communication to the approaching vessel. Two shuttles were hidden on Phobos. Also, the *Calypso* had nestled up next to Phobos in an attempt to hide its presence from the approaching vessel. Both the *Calypso* and Phobos looked like irregularly shaped moons.

"Maybe now is the time to send a message?" Kwame asked.

Gus answered, "I agree. If we send something now, it might seem completely innocent. They might conclude our radar was not operational, and we only had telescopes. Let's ask them to identify themselves."

"Unidentified vessel, this is the Mars Musk Colony, please identify yourself, over," Ginger asked.

"*Nautilus,* we are. Earth we are from, over" came the reply.

Ginger was listening intently. "That doesn't sound like English speakers. This might be some sort of deception."

"Wait, my boyfriend, Carlo, told me that when they landed on the Earth in the other universe, they spoke a strange version of English. He would joke about it and start talking to me in that exact manner. Let me ask them about their origin," Cassia said.

"*Nautilus*, this is the Mars colony. Are you from this universe or another one? On our way here, we accidentally entered a different universe when passing through a wrinkle. Do you recall if your Earth received a visit from Earth people from a different universe? Do you know a Dr. Levvi Markanen? Over."

"Yes, a different universe from we are. Dr. Levvi Markanen know we. *Calypso* from you are?"

"Yes, we know him, and yes, we are from that vessel. We won't answer any other questions over the radio. We don't know who might be listening. Why have you come to this universe?" Cassia asked.

"Help to you come we. Understood we technology ours is advanced more. Help you will we. Survivors will we get."

"Do I understand they will help us get survivors and will help defend us against the AOW?" Kwame asked.

"I believe so," Cassia answered.

"*Nautilus*, we copy. Thank you for your offers of help. We certainly can gather survivors and bring them to our vessel more quickly with your help. I suggest we coordinate that here at the Mars colony. The less we say over the radio, the better. Can you land near these coordinates and discuss this with us? Over," Cassia said.

"Affirmative, *Calypso*. Our commanding officer and some others will descend to the surface. We will endeavor to speak Old English with you," came the reply.

* * *

"I am Captain Tennyson Franklin of the Unified Air and Space Forces. It is a pleasure to meet you, Mr. Olajuwon. I wanted to wait until we met in person to tell you of some most distressing news. After your shuttle left our Earth, maybe three months later, another spacecraft landed on our world. These people identified themselves as representatives of the AOW. I believe these are the people who destroyed your Earth and have been harassing the *Calypso*. They told us they were looking for your ship and a girl who was kidnapped by the *Calypso* crew against her will. This girl they referred to as Terra Nova Winstead. We had been told some things concerning these people, so we knew they were dangerous. We told them we had not seen another vessel nor did we know of this girl. They were most pushy and aggressive. They were vile people, very greedy, selfish, and we believe they are willing to break any norms or laws to achieve their aims. We can sense these things. Such aberrant personality types were eliminated from our genome over one hundred years ago. I believe they may be on the trail of the *Calypso*. We will help you gather up your people on Earth so that you can leave as soon as possible before they catch up to you. We will remain behind to slow them down, and we will destroy them. Our technology is at least the equal of theirs, and we have excellent tactics."

Kwame thought for a moment, then said, "Thank you for your assistance with these dangerous people. We would be most appreciative of your assistance in gathering up survivors on the Earth. Are you sure you can fight these people? I don't want you to get into something you can't win."

"We would win, I assure you. We had a look at their ship. There are key pieces of technology that they have not yet developed. We have technology that can make us completely undetectable to radar, infrared, or visual technology. We will destroy them for you. We know what they are capable of. This would send a message to them not to enter our universe again."

"If possible, we would like you to, somehow, capture some of these people so that they can be interrogated. Perhaps you can learn where their home world is located and other useful pieces of information. Therefore, I advise you to disable their ship and take prisoners. This would be most helpful to us. I think you have what we call EPC tech. If you have matching sets, you could communicate with us and let us know of anything you learn about them. Of course, this means you would have to enter our universe again, or at least I think that, to send us a message."

"We will consider this course of action," Captain Franklin said.

PART FOUR

Earth Rescue

CHAPTER 38

Earth Rescue

FOR THE NEXT seventy-one days, the *Calypso* informed three survivor groups on Earth that a rescue was on the way. They asked each group to try to contact other groups and let them know they would soon be rescued. One group that they were not able to reach was the group in the Denver suburbs. They realized they would have to hunt for some of the groups once they landed. Terra Nova had informed them that survivors in the Denver suburbs included both sets of Terra's grandparents. An attempt was going to be made to rescue them based on the origin coordinates of the communication they had previously made with the Mars colony. By the time they established Earth orbit, the *Nautilus* and *Calypso* had coordinated locations that each would be responsible for evacuating survivors from.

Levvi Markanen piloted his shuttle to a landing within two kilometers of the last known origin point of radio communication between Mars and the Denver survivor group. With him on the shuttle was Captain Sandra Simmons, who insisted on going to the surface to help locate Terra's family members.

“I’ll take Ms. Stetson, Buck Stilson, and Kwame with me to locate the survivors,” Captain Simmons told Levvi, who was required to remain with the shuttle.

Captain Simmons and the others were dressed in winter clothing. Both the season and the obscured atmosphere from suspended debris made for very cold conditions.

“Ms. Stetson, you lead the way. You have our GPS,” Captain Simmons said.

As they walked to the coordinates, they saw a person scurry from a pile of debris to the inside of a long, two-story building that was mostly intact.

“I see movement—a person. I think we should follow that person into that building,” Anna Stetson said.

“Copy that. Everyone, be on guard. These people may not figure out who we are. They may see us as a threat. I’m going to announce our presence on the loudspeaker. Perhaps that will let them know we are not a threat,” the captain said.

“Attention, please. I am a captain in the Air and Space Forces. We are here to evacuate survivors to the Earth colonies on an exoplanet in the Cervantes star system. We are here to rescue you. Please make yourself known to us.”

Inside the building, a group of 121 survivors huddled in the basement. They had heard the loudspeaker message but were unwilling to show themselves. The *Calypso* crew did not know it, but this group had lost four of their members in an ambush with a group of marauders a few days earlier, who had convinced them they were there to help them.

"Ed, I think we should go upstairs and meet with them. I have a feeling they are somehow connected to Sam, Riley, and Terra. They said they will take us to an exoplanet in some other star system. Ed, it has to be for real this time," Kelly, Terra's maternal grandmother, told her husband, Edward.

"John, what do you think?" Edward asked John Winstead, Terra's paternal grandfather (Sam's father).

"I'm with your wife. I think there is practically no way they are not here except to rescue us. Why would they tell us they are here to take us to an exoplanet? Their cover story is so bizarre, it has to be real this time!"

"OK, but only I will go up and approach them," Edward said.

"No way, I am going up with you," Anne, Riley's sister, said. "Mother, stay here and watch Brittaney."

"I'm coming also," John said. "I think this is the real deal. We are finally being rescued. I'm willing to take the chance. We'll starve to death here eventually!"

"Coming up! Don't shoot. We are unarmed. Our people down here are armed. If you hurt us, they will attack you. We have very little food. We have nothing of value for you. Our radio no longer works. We have nothing! Coming up!" Ed shouted up the stairs.

Edward, Anne, and John walked up the stairs and walked toward the four people at the far end of the first floor of the warehouse.

"I am Captain Sandra Simmons, commanding officer of the starship FBS *Calypso*. We have established an Earth colony on a planet we call Terra Nova after our leader, Terra Nova Winstead. We learned of the destruction of the Earth from the Mars colony. We have returned specifically to rescue Mars and Earth survivors and evacuate them to Terra Nova. Can you identify yourselves, please? How many survivors are there at this location?"

"Daddy, Riley, Sam, and Terra are alive! The ship was not destroyed! They know Terra!" Anne said excitedly to her father, Edward.

Edward nodded, then responded, "I am Edward Singer. There are over one hundred survivors here. My daughter, Riley, was on board the *Calypso*, as was my son-in-law, Sam, and their daughter, Terra. Are you really from the *Calypso*? How are they? How are Sam, Riley, and Terra?

Captain Simmons smiled and answered, "They are fine, Edward. You have new grandchildren and great-grandchildren. Terra's story is a bit complicated, but she is doing well. We need all of you to gather your belongings and follow us to one of the *Calypso* shuttles that we piloted to a landing site not far from here."

Edward, John, and Anne shook Captain Simmon's hand as they all broke down from the emotion of this highly unexpected moment.

When they reached the shuttle, the captain introduced them to Levvi and the other crew members on board.

Captain Simmons asked, "Edward, John, I'm told you are aware of other groups in this vicinity. Do you know their locations? We are trying to rescue as many as possible. Most of you will be going into hibernation sometime soon after our arrival back on the *Calypso*. Those with children under the age of eighteen will stay awake until their children reach eighteen, which is the safe age for being put into hibernation, unless there are other underlying health issues, of course."

"Well, I can tell you that one group is in Fort Collins. I assume your spacecraft can take us to that location? The other group is near Grand Junction. This is also, obviously, much farther than we could walk to. I think the best we can do is to go to those locations and somehow announce our presence. We haven't seen any other groups of people near here. There may be a few people here and there. I know that people are getting hungry. The supplies of food are getting much harder to locate. We've tried to grow a few things, but the weather just isn't warm enough, long enough, to grow much. Your timing couldn't be much better. We were getting very desperate. Again, thank you, thank you so much," Edward said, breaking down again with tears of gratitude. "We've been through so much, you can't imagine how hard it has been. My daughter, Anne, lost her husband when the lunar crust impacted north of Denver. Later, she lost her younger child. Everyone has lost so much." Edward began crying and collapsed on the ground.

“Levvi, get Dr. Le Cesne up here. Edward needs medical attention,” the captain said.

* * *

“He’ll be fine. The roller coaster of all this caused him to be overcome by emotions. He has been the leader of this group of survivors, and that burden, under these extremely trying circumstances, has caused him to develop post-traumatic stress syndrome. He’ll need time to process this all. I’ll speak with him and give him with some medications to help with his anxiety. It will help him sleep better, which I am sure has been difficult for him.”

“Thank you, Dr. Le Cesne. We’ll let him rest. John and some of you others, we’ll rely on you to help guide us to these other groups he spoke of. Everyone, let’s get you strapped in. We’ll be flying the shuttle north to Fort Collins. Our shuttle can fly like a helicopter to reach nearby locations. Everything is going to be OK, everyone,” Captain Simmons said.

CHAPTER 39

Gathering Completed

FOR THE NEXT three months, the landing parties from both ships were constantly descending, locating, loading, and returning to the *Calypso*. In the end, over twenty-seven thousand survivors from Mars, Europa, and the Earth were located and ferried up the *Calypso*. Most were now in hibernation. A number of people required medical attention before they could be safely put into hibernation.

On the last day on Earth, the last shuttle, piloted by Cassia, was hunting for survivors in Washington, DC. Several dozen were located, including two congressmen and three individuals who were cabinet secretaries. As they lifted off for the final time, Cassia breathed a sigh of relief.

"This has been so difficult, both physically and emotionally. To see the emotional scars—that is something I didn't think would affect me as much as it did. Everything in this world is pretty much destroyed. I know Terra said that our descendants would someday return and reclaim this world, but right now that just seems so impossible," Cassia said to Levvi, who was with her on this last flight.

"I also can't imagine what they have gone through. My takeaway from all of this is that the AOW is just so cruel and heartless to kill so many and then have the survivors clawing for survival for years. No one has said this, that I have heard, but I wouldn't blame these people if they wanted revenge at some point. The AOW are absolutely terrible people. They are evil," Levvi said.

"Their day will come. They will eventually pay for what they have done. Without any sort of punishment, they may never feel remorse for their actions," Cassia said. "Let's go home."

CHAPTER 40

Get Out of Dodge Now!

"CAPTAIN SIMMONS! THIS is Captain Franklin. We left an autonomous detection system out near your Europa colony. It pinged an alert to us moments ago. I'm afraid this might be the AOW. We are breaking orbit to engage this vessel. It will be weeks before it reaches us, we will let you know the outcome. Advise you leave forthwith and at top speed. Good luck!"

"Happy hunting, *Nautilus.* Godspeed," Captain Simmons responded.

"Ms. Stetson, set course for Terra Nova, maximum speed. Give us four solid rocket motors as a boost!"

"Aye, aye, captain," Anna responded.

"This is the captain. We will be leaving Earth orbit sooner than planned. Brace yourselves for acceleration as the solid rockets push us out of orbit. Our friends from the *Nautilus* have warned us of an approaching vessel way out near Europa. It is likely the AOW. We do not want to hang around here any longer. They are going to delay and disable that vessel for us."

The ship experienced a modest acceleration as the solid rockets ignited and the *Calypso* moved out of orbit.

* * *

Captain Franklin engaged the cloaking system as their vessel entered an orbit around Jupiter. He planned to hit the nozzle of the AOW ship with his rail guns. He would offer the disabled ship the chance to surrender. This would allow them to take prisoners and have them interrogated in accordance with the wishes of the *Calypso*.

"Captain, I have a visual on the ship. They are using the Jovian atmosphere to slow down as you speculated. I have the targeting system up and have a lock on the AOW ship. Do we hail them first to ascertain who they are?" Ensign Kristoff asked.

"Hail them. Ask for identification. Refer to them as the 'AOW ship'. They may lie about who they are. Let's give them the chance to deny it or miss that we refer to them as the 'AOW ship'. Their answer will reveal their intent to me and what I choose to do—fire on them or let them continue on their way. There is no reason for any ship to be in this solar system, given that this Earth was destroyed."

"AOW ship, identify yourself and your intentions. This is the Earth ship *Nautilus,* over."

"Earth ship, this is the AOW ship *Invincible*. Prepare to be boarded!"

"*Invincible*, I was just about to inform you that we would be boarding you! After we disable your vessel, we will be boarding it. Unless, of course, you refuse to surrender, then we will destroy you!"

"Ensign, fire on their nozzles."

"Aye, aye, Captain. The AOW vessel has been struck by our rail gun. Their nozzle is fractured. I guess they were not invincible. Should we demand their surrender now?"

"Yes, Ensign, I shall now do that. *Invincible*, we demand your surrender, otherwise we shall keep firing until nothing remains of your ship. Your choice. You have ten seconds to respond. Failure to respond shall be taken as refusal."

"Don't fire! We surrender!"

"Good. Supply instructions on how and where we will board your ship, over."

CHAPTER 41

The Long Sleep

"CAPTAIN SIMMONS, I want to thank you again for everything you have done for us. It meant so much to us to be able to speak with Terra and her husband, Hatcher, and to see the new additions to our family. I hope no one is expecting me to start having children again! I'm too old," Edward said to the captain.

"Edward, I am so happy we could arrange to have Terra and her family speak with you. The next time you see them, you will be in their presence. That is going to be so exciting! I'm told you checked out OK to enter hibernation. Trust me, that is the best way to travel this great distance. This will be the second, and I hope last time, that I will be on this leg of the journey to Terra Nova, although it is much shorter than we originally thought it would be! Sweet dreams! See you in 'the morning'!" Captain Simmons said.

"Come, Kelly, it is time to go to bed. At least, I don't need to worry about you taking all the covers!" Edward joked.

"I have a comeback for that, you know! But I think I'll spare everyone from 'too much information'!" Kelly answered.

“All right, you two, let’s get you two put to bed,” Dr. Le Cesne said.

“We’ll see you when you both wake up. We’ll be orbiting Terra Nova by then. This is so exciting. This little girl will be a big girl when you wake!” Anne said, holding her daughter, Brittaney, with an arm around her shoulder.

“Night, night, Grandpa and Grandma. Sleep tight! Don’t let the bedbugs bite,” Brittaney said.

CHAPTER 42

Interrogation

"CAPTAIN STILES, AGAIN, I ask you. Where is your home planet located?" Tag Fortman asked the *Invincible* captain.

"I am not telling you anything."

"Captain, are you familiar with the devices that were implanted into the heads of two crew members of the *Centauri Observer*?"

"I don't know what you are talking about!"

"Really? Well, it is this device," he said, holding up the device removed from Tariq and Shandra. "It allows two-way communication between the individual who has the implanted device and the person with the control unit. It is an EPC device. I know you know what that is, your ship had such communication technology. It can also deliver a powerful pain stimulus directly into the brain's pain center. I'm told that there is no other type of pain that compares to this pain. It was so painful, in fact, that it caused cardiac arrest in the young woman who had this implanted into her head by you monsters. We found more of these on your ship. Apparently, a number of your crew have these implanted devices to help them 'remain compliant.' I believe those were the words your ship's doctor used before he died from the pain they delivered. It took him fifteen minutes to expire from the pain. Really not a long time, but I'm sure it felt like an eternity to him. Still don't want to cooperate? We can put this into your head, but it is likely to be a permanent fixture in your head if we do it. You see, we made an improvement on it. We put a small explosive device in it. Enough to sever some arteries, but not likely enough to kill you, but you will be severely disabled and an idiot, having lost vital functions controlling your body, like bladder control, movement, speech, etc. We tried it out on the good doctor—worked like a charm! Well, for fifteen minutes, then he expired. But we think we've got it worked out now, to where it would just cause irreparable brain damage. We'll never know until we test it on you! So do you want to tell us where your home world is located, and *all* the other things we want to know?"

Captain Stiles thought for a moment, and then he started to cry. "OK, I'll tell you everything you want to know, just tell me you won't put that damn thing into my head!"

"I think you made the correct choice, captain, though I am disappointed that we didn't get to try it out with the improvement!"

* * *

"Well, Doctor, looks like your captain bought into our little ruse, though I do still want to try this out on some of your people. How about you? This would help advance medical science. You are a man of science—would you like to volunteer? You see, I know you were a pioneer in developing this tech. You could learn so much by trying it out yourself! We spoke with Tariq, he was insistent that you would be willing, and excited to try it out personally. He and Shandra were engaged, you see. He took her death quite hard. Shall we?"

"No, no! I'll cooperate! I was only following orders when I put it into them! Please, I was only following orders! You must understand, you are a military man also!"

"Yes, my good doctor, I am a military man—one who specializes in interrogations. But I am not a monster like you. I think that you could be a good visual example of the power of this device for the others in your crew. They could see its effects on you, and they will also likely become 'compliant.' Don't worry, I'm not much into needless suffering. If you are compliant, you never need to experience the pain. What do you say? Let's get you prepped for surgery!" Tag said, smiling.

* * *

"Captain Franklin, to summarize, the AOW home world is Kepler 452b, which orbits the star also called Kepler 452b. This star is about 1,400 light-years from Earth, or 1,350 light-years from Cervantes star. Did I get that correct?" Captain Simmons asked.

"Yes. Other important pieces of information are in the report I scanned and sent to you. One of the most important pieces of information is that the AOW does not possess cloaking technology. We will share this technology with you when we put it together into a format that will be easily understood. It would be most interesting to approach their world using cloaking tech, land on it, and gain even more useful information."

"I agree. It will have to occur at some point in the future. My job now is to get all these folks to their new home first. Many thanks again for all your assistance."

CHAPTER 43

Evening Up the Odds

"CAPTAIN, THESE DRAWINGS and specs Captain Franklin sent are rather complete, and in my view, not impossible to actually implement into a working system for the *Calypso*," Timeye said. "I think it won't hurt anything to try and make a working system for our vessel."

"What do you think, Stuart and Dr. Markanen? Can we do it? Should we try? I mean, we have almost thirty thousand people on board. If the AOW were to take us out, it would be a serious blow to our efforts to reach a critical mass of population on Terra Nova. Not to mention a great tragedy," Captain Simmons said.

"It's worth a try, there is not much downside, and the upside is obviously very large if we are able to survive an encounter with a superior enemy," Levvi said, and Stuart nodded in agreement.

"OK, let's do it. You three have my full support," Captain Simmons said.

* * *

"The basic premise behind the cloaking is that another visual image can be forced back toward the enemy vessel, essentially showing empty space in place of a visual image of the vessel they are looking for. Of course, radar pulses have to be absorbed to hide the ship from radar searches, but we know how to do that. That will be the second phase of our efforts," Stuart explained.

"What about gravimetric searches?" Timeye asked.

"We can mask the curvature of space created by the *Calypso* by sending out anti-gravity particles to foil gravimetric searches," Levvi answered.

"Let's get to work!" Stuart said.

* * *

Six months later, the cloaking system was completed. A test of the system, with a shuttle viewing the *Calypso*, proved successful. Only empty space was seen from the shuttle when the system was activated. The other two phases of the ship protection system, or SPS, were going to take about a year to complete, well before they reached Terra Nova.

CHAPTER 44

Showdown in the Heavens

"CAPTAIN, I'VE GOT a large reading on the differential gravitometer. I don't think it is a natural object," Anna Stetson announced to the captain and other assembled crew members.

"Engage the SPS, stat," Captain Simmons shouted.

"Engaged, Captain."

This was the second time in the last two wrinkle transits that something had been detected, something that resulted in a scrambled effort to employ the SPS. The first time was a system error, not a threat from an AOW ship.

"Timeye, what do you see on our passive sensors? Specifically, what do you see on infrared?" the captain asked.

"I see a large ship outline, ion discharge. It's definitely a ship. The other times we transited this wrinkle we never detected any ship or other manmade object. Without hailing them, we won't know for certain, but who else could they be if not the AOW?"

Stuart had just entered the guidance and control room. He had listened, and now he spoke. "I think we should fire on their nozzle and try to disable them. Then we could speak to them once they know they are helpless."

"I agree, captain," Timeye chipped in.

Captain Simmons called down to Buck Stilson, who was manning the ship defense console in the observatory room.

"Buck, fire the rail gun at their nozzle!" the captain, now operating on adrenaline, shouted into the com mic.

"Copy that, captain."

The rail gun fired two rounds at the vessel; the third shot destroyed the nozzle.

"Nozzle destroyed, captain," Buck reported.

"Ms. Stetson, hail the AOW ship, demand their surrender."

"Aye, aye, captain. AOW ship, we demand your surrender. Do not fire, or we will continue firing on you until your hull is breached!"

"We surrender! We surrender! Do not fire on us!" came the response from the AOW ship.

"Captain, they have acknowledged our demand and have surrendered," Anna Stetson announced.

"Copy that, Ms. Stetson."

Just then, a loud bang was heard ringing through the *Calypso*.

"What was that, captain?" Stuart asked.

"Did they just fire on us?" Captain Simmons asked to know one in particular.

"I believe so, captain," Timeye answered.

"Buck, return fire! Finish them off!" Captain Simmons shouted to Buck through the com.

"Copy that. Returning fire!"

"Captain, they have been struck multiple times, recommend hailing them again to determine if they are still operational and a threat to us," Buck responded.

"Ms. Stetson, hail them again."

"AOW vessel, are you still operational? Again, we demand your surrender!"

There was no response to repeated attempts to reach them.

"Observatory room, can you visually see their ship? What do you see?" the captain asked.

"We see two large sections of a ship, with numerous smaller pieces. They have been destroyed," came the response from the observatory room.

"Captain, should we check for survivors with a shuttle?" Timeye asked.

"Negative, we can't afford to lose the velocity we have attained. Finding survivors is very unlikely." The captain now spoke to the crew, "Attention, this is the captain, all sections report damage and casualties."

The resulting reports confirmed there were no casualties and no damage. The *Calypso* had faced down a superior opponent and destroyed them, achieving a great victory. The return voyage would now continue undeterred. Terra Nova was just six years away.

CHAPTER 45

Homecoming

"CAPTAIN, I HAVE an EPC message for you from Sam Winstead," the com engineer on duty said.

"Copy that, read it."

"Captain Simmons, when you come within conventional communication range, which we define as less than a thirty-second delay, we will be sending you an orientation video, which we would like to be played for all new citizens of Terra Nova. Additionally, we would like your crew to watch it also. It will be in three one-hour sections, dealing with a description of Terra Nova in part one, including unique hazards to human life, a section dealing with colony life and governance in part two, and a section dealing with our mission on Terra Nova, the development of a virtuous and intelligent species on the planet. Lastly, we will have a live set of discussions when the com lag is less than six seconds. These discussions will deal with the unloading of passengers and the disposition of *Calypso* assets, shuttles, etc. We want to put the ship into an elliptical orbit around the Cervantes star and Terra Nova. We will need to have a rotating skeleton crew on board at all times to maintain and control the ship. Let us know of the availability of you and your officers for these discussions. We look forward to seeing you and all the new arrivals within the next two months. Terra Nova coms out."

* * *

Two months later, the elliptical orbit had been established, and shuttles began loading up passengers and ferrying them to the surface. Given the very large number of people to be brought from the ship to the surface, it was going to require that all three shuttles from the *Calypso* and the three that The Mother's people had constructed be constantly employed in the effort. Hatcher, Gunner, and some of The Mother's people were busy with the effort of flying the shuttles down and back up to the ship. The effort took six months, and there were a couple of close calls as the people and equipment were being worked to their breaking point. At the end of the six months, the unloading was completed. All six shuttles needed extensive preventative maintenance given their high usage. This effort took the next year to complete.

* * *

"Mom, Dad, Anne! I never thought I would see you again!" Riley said through tears.

Sherri, Hatcher's mother, ran up to her father, John Moore, and hugged him.

"Oh, Daddy, it is so good to see you again!" Sherri said as her husband, Nathan, stood at her side, saying nothing.

"Who are these young people?" John Moore asked his grandson's wife, Terra.

"This is Willie, and this is Connie. They are your great-grandchildren. You have another great-grandchild, Timothy, he is one year old. He is being watched at home. It was his naptime. And this is Jacob, my son through Ammon, my deceased second husband," Terra explained to John.

"John, it is so good to see you again! I wish your wife were here with us," Nathan, Hatcher's father, said to his father-in-law.

"I know, I know. I think about her every day. I wish she were here also. "Now, where is Hatcher?" John asked.

"Well, he is a shuttle pilot, so he is very busy. We'll see him tomorrow. This is his wife, Taylor. She is a native Terran. This is her son, Zach, and her daughter, Charlene or Charlie."

"I see. Well, it is nice to meet you, Taylor, Zach, and little Charlie," John said as he picked up Charlie and gave her a hug.

"We are honored to meet you, Mr. Moore," Taylor said.

"Come, let us show you where you will be living," Sam said as he, Riley, Terra, Taylor, Nathan, Sherri, and assorted children began walking off in the direction of the new dwellings that were on the south side of Benston.

CHAPTER 46

Reacquainting

LEVVI WAS ONE of the last to come down to the surface due to his duties as a shuttle pilot. The shuttle he was on when he finally came to the surface was piloted by Hatcher. They had met several times as both were involved in ferrying people to the surface and had crossed paths. The first time was on the *Calypso* when Hatcher had stayed overnight after returning from a late-afternoon trip to the surface and back. Levvi knew who Hatcher was, but he had not said anything about his recollections of Terra in a previous life. Levvi did not know what to say.

Levvi went to Stuart to ask him what his thoughts were on how, or even if, it was a good idea to meet Terra and speak to her about what he remembered.

* * *

"Well, Levvi, I don't see a downside to speaking to her about your remembrances. We believe in multiple lives and progression through each life. She teaches that. You certainly fit the description, physically, of Ammon Khan."

"How would I 'seek an audience' with her?"

"It is my understanding that she was, at one time, meeting with Ammon's people once a week to hear 'issues,' disputes, things like that. I believe Taylor was involved as a scribe to write down what was heard and decided. I know that Connie, her daughter, has been picking up these duties, or so I've heard. Perhaps I could ask Taylor to come and talk to you. Maybe you could tell her your story, and she could tell Terra. I'm certain that Terra would listen to you, however. She is a very open and kind person."

* * *

"Taylor, may I speak with you? You might not remember me. My name is Stuart Carroll. Perhaps you've heard about the experiences that the *Calypso* had on the way to Mars. At one point, we entered a wrinkle and ended up in another universe, or spacetime. While in that other universe, which had another Earth that diverged or split from my Earth's timeline, we went to that Earth to get help getting back to this universe. While there, we met a very interesting man, a physicist like me. His name is Levvi Markanen. He is the very tall man you may have seen walking around here in Benston. Have you seen him?"

"Yes, I have. He looks so much like my deceased husband, Ammon Khan. Terra made the same observation. We would like to meet him but haven't had the chance to do so. Could you introduce us to him?"

"Yes, I would be happy to do so. He is staying with me at my house. Can you come by my house tomorrow morning sometime? How about 10:00 a.m. tomorrow? You can find my address in the colony listing."

"I believe neither of us has any obligations tomorrow morning. We can bring our kids to their grandparents' house and drop them off, then come over to your house."

"Good! I'll see you then."

* * *

"Terra, what could this Levvi Markanen have to tell us that is so very interesting?" Taylor asked.

"I think I know, but I'll wait to say, if you don't mind."

* * *

Stuart came to the door when Terra and Taylor knocked.

"Terra, Taylor, please come in. Levvi, can you come out here into the kitchen? There are two ladies I'd like you to meet," Stuart shouted back to the bedroom where Levvi was.

"Sure, just a minute."

As Levvi walked briskly into the kitchen, he slowed down and stopped when he saw the two women. He started to say something, then stopped. Terra walked up to him and hugged him, kissed him on the lips, and then started crying while holding him tightly. Taylor, who must have figured out what this meant, rushed over to him, hugged him, kissed him, and then began crying as well.

"Now, here I thought this was going to be impossible to explain, but it seems that both of you know something of what I was going to explain. I was told Terra could read minds, but I wasn't sure if I believed that. I do now!" Levvi said with a chuckle as he tried to wipe away a couple of tears that he was shedding.

"My love, you are back! What are we to do now?" Terra asked.

"What will we tell Hatcher? How will he take this?" Taylor asked.

"I've already talked to him and told him that Ammon is back. He was not sure how he felt. He and Ammon became good friends, and his death hurt Hatcher like it did all of us," Terra said.

"Well, that was easier than I expected," Stuart said.

CHAPTER 47

Renewing the Vows

LEVVI, TERRA, TAYLOR, and Hatcher got together the next day at dinner. An awkward discussion ensued.

"I understand that Levvi believes he was Ammon, and maybe he was. But my issue is that he is now a different person. He grew up on a different Earth, he has a different occupation, and he has had different experiences. It is not a straightforward assumption that he can pick up where Ammon was when he died several years ago. I think he, and we, should give all of us some time to process this. Maybe he wants to 'play the field' before he commits to the relationship we all had before Ammon died. Remember, Ammon and I were friends. I see Levvi, and I think I am seeing Ammon, but it is not him. He is a different person. He is Levvi, not Ammon," Hatcher said.

"That's true. He may not want the same arrangement. Maybe he thinks it wouldn't work. I think the culture on his Earth was like it was on yours. You did not have a culture of multiple husbands and wives," Taylor said.

Terra listened to this, then said, "Well, he is right here. Let's ask him. Levvi, what do you want to do?

Could you live with an arrangement where the four of us share each other? Do you know that Taylor and I also love each other and have an intimate relationship as well? Are you OK with that as well?" Terra asked.

"I'm OK with any or all of this. I remember most of what happened before in my previous lifetime. I remember Ammon, I remember both of you ladies, and I remember Hatcher. I know this is not typical, but I do remember. I also remember Eve told me I would eventually remember these things," Levvi said. "But I can wait a while, if that helps everyone adjust to this unusual situation."

"Maybe it would be best if Taylor and I 'move aside' and let you and Terra become acquainted again. Taylor, what do you want to do?" Hatcher asked.

"Perhaps it is best if he gets to know each of us, one at a time. Hatcher and I can move back to Benston. Terra and Levvi can stay at their home in Ammonsberg. But I do want to have my time to become acquainted with Levvi when Terra is done becoming acquainted again."

CHAPTER 48

You Can't Always Have What You Want

LEVVI WAS HELPING Hatcher and Taylor pack things. They were loading it into a horse-drawn flatbed wagon. Terra was watching their children.

"Can you help me with this burlap sack, Levvi?" Taylor asked.

"I can just get it myself. Why don't you go spend time with Terra and the children?"

"Well, I think that might be it. We'll leave some of the bigger things here since you and Terra have the 'palace'!" Hatcher said.

"Hatcher, what were some of the things that you liked about Ammon?" Levvi asked.

Hatcher paused, then answered. "Well, there were a number of things. First, he was kind, especially to Terra and the children, but also to me and Taylor. He really did care about his family. He spent time with them. He did this even though he was the Khan. That position brought a lot of responsibility and took up much of his time. But I also liked that he enjoyed hunting. That is where we really got to know each other. He wasn't above getting his hands dirty."

"I have never hunted. I assume that is how the people in Benston and Ammonsberg get most of their protein?"

"Yes. Before we arrived, very few people here—in this area anyway—had engaged in growing food. Just a few people here and there. A few years back, we went to a far distant empire presided over by a man who took the Earth name Genghis Khan or GK. They had successfully developed agriculture. The Terrans around here have picked up on this rather quickly, and they do a great job at it, though they still hunt for some of their food. So far, we haven't had too much luck domesticating the wild version of cattle, which we are calling aurochs, but we are still trying. On Earth, these were the wild ancestors of cattle. We give the same name to many things here that we had on Earth, FYI. Well, I think we are done, we should be going. It'll get dark soon. Let me know when you have some time. I'll teach you how to hunt this time around," Hatcher said, laughing.

* * *

Levvi sat quietly and listened as Jacob spoke.

"I never knew my Dad, he died before I started remembering things. I think I wasn't even one year old. Isn't that right, Mom?" Jacob asked. Jacob had just turned seventeen a month earlier.

"That's right, honey," Terra answered. "He looked a lot like Levvi, if you ever wanted to know what he looked like."

"Is that where I got my height from?" Jacob asked. "I mean, I didn't get it from my petite little mother!" Jacob was nearly 209 centimeters, and he was still growing, though his growth was finally slowing down. He would likely end up as tall as Ammon or Levvi.

"No, honey, you definitely did not get your height from me!" Terra answered with a smile.

"So Hatcher was really the man you knew as your father?" Levvi asked.

"Yes. Hatcher is my father, at least in the way I think of fathers. He taught me a great many things. He's even taken me up in the shuttle, and he is teaching me to fly it. I guess it helped that my mom force-fed me a lot of physics! It has helped me understand how the shuttle works. I'd like to go up to the *Calypso* someday on the rotating crews. It would be so cool if we could go to another planet on the starship. I don't think we have any plans to do that," Jacob said sadly.

"Well, not right now, but maybe a time will come when it will be necessary," Levvi responded.

* * *

Later that day, Terra left little Timmy in the care of his older brother, Jacob. She and Levvi went for a walk down to the seashore.

"Why is everyone staring at me?" Levvi asked.

"Probably two reasons. First, you are very tall, even for this group of Terrans. Secondly, you look a lot like Ammon, the old Khan, their leader. I imagine they are really not sure what to make of you. They see me and you together and they think Ammon is back somehow, which I guess he is, in a way."

"I don't know if all this is such a good idea," Levvi said as he looked out at the ocean, deep in thought.

"What do you mean?" Terra asked, suddenly concerned.

"I mean," he paused. "I mean, trying to reconnect to everything. I'm different. I am not Ammon Khan. I'm me, Levvi Markanen. I don't have the right to come here and start living with somebody else's wife. It's not right. I have to be true to myself and what I think is right. Terra, I love you. I want you. Hell, everybody wants you. I can't be with you. I have to find my own life, not Ammon Khan's life. I'm sorry."

Terra didn't respond. Instead, she softly cried. She knew he was right, but it still hurt so badly. She had her love back, but he wasn't the same. He wasn't going to stay.

"Let's get back to your house. It's dark. I'll stay the night, then I'll move back in with Stuart."

PART FIVE

Building a Better World

CHAPTER 49

Nutrition and Trade

THE CLIMATE AROUND Benston allowed for many crops to be grown, but the yields on some crops were not enough to meet the dietary needs of everyone. In particular, citrus fruits were lacking. What could be grown in greenhouses was not enough for everyone. Additionally, the last two years had seen a cooling of the climate, and this had caused crop yields to be below average. A meeting of the supply council at the end of the growing season was convened to discuss these issues.

"We're going to have to use our reserves to make it through the winter. We won't starve, but we're going to have to hunt and fish more. The Terrans have boats, and we need them to be out on the water fishing every day and week. We also need to hunt and salt away meat from land mammals," President Stanton said.

"We're all in agreement with you. Perhaps we can trade with groups of Terrans far and wide. Maybe we can go back to GK's kingdom and see what we can trade with them," Hatcher said.

"That's a good idea," President Stanton said. "We should get a trade mission planned right away. Gunner, can you draw up plans of who will go and what we might have to trade with them?"

"I'm on it," he responded. "Though I do have one concern about using the shuttles for this trade mission. I know our reserves of hydrazine are low. We can use the spinners. They draw power from the thorium reactors, so no problem there. But if we ever had a pressing need to use the rocket engines, we would be limited. Recall we had to expedite our return from the high plains when the pandemic broke out. Something could happen again, and we may need to use the shuttle's rocket engines. In my opinion, this is a supply issue just like food. What can we do about this?"

"There are thousands of liters of liquid rocket fuel on the *Calypso.* While we still have enough rocket fuel to go up to the *Calypso*, we should do so. We need to go get the equipment that is needed to produce this fuel. As you all are aware, the *Calypso* has a small manufacturing operation that can produce it. We don't. We should set that up down here," Riley's sister Anne, said. Anne was a chemist.

"I agree with Anne," Peter said.

President Stanton had listened intently to the conversations. He stood and said, "I want Peter and Anne to put a plan together for production of liquid rocket fuels here on Terra Nova. As I said a moment ago, Gunner will put a plan together for a trade mission to GK's kingdom. Terra, do you have anything to add to this discussion?"

"No, I think you got it covered, Mr. President."

CHAPTER 50

Trade Mission

GUNNER HAD DEVELOPED a plan for the trade mission to the kingdom of Genghis Khan. He had a roster of people who were going, and he had come up with a number of items he believed they could trade for fruits and vegetables.

"Connie, I want you to come on this trade mission because you can help us in ways that Terra has helped us. We believe that GK is in his mid to late twenties. Your mother had a friendship with him. We are hopeful that you can become friends with him as well," Gunner explained.

"Gunner, what kind of things do we have to trade with them?" Connie asked.

"We have salted salmon and turkey. I never saw those things during the time I was there. We also have Terrans who have boat-building skills. We are thinking that we can show them how to build ships and navigate them on the open ocean. This could lead to more trade in the future. I never saw or heard that they engaged in sea travel. I think they just don't know how to build ocean-going boats. We've got maps of the hemispheres that they could find very useful. Many of the things we hope to trade are technology items that we think will enhance trade and communication among Terran groups. This has all been approved by your mother."

"I think it sounds exciting! I've never really been anywhere in my life. It will be an adventure for sure! Thank you for letting me come with you."

"Well, we kind of need you, Connie. You are very important to our colony and what we are trying to do on this world."

* * *

The trade mission left on a Monday morning. In total, there were thirty Earth people and another thirty Terrans. The shuttle was packed with trade items. The foodstuffs were all salted and were packed in boxes built by the Terrans. Hatcher and Gunner were piloting the shuttle.

The flight over the Terran Gulf was smooth. The air planet wide was cooler, and there seemed to be less wind. They flew the entire day and stopped for the night about twenty kilometers inland. The air was much cooler this time, Gunner and Hatcher noted.

Things just don't feel right. The land and plants seemed to know that something was wrong. The sun did not shine as brightly. The whole planet knew it, Hatcher thought.

At dawn the next day, they were greeted with a strange sight. A group of Terrans approached the shuttle and asked for help. Their biggest need was food. The fish they used to catch had either died or gone somewhere else.

"We can give them some salted salmon. Some other food items as well," Connie said to Gunner.

"For sure," Gunner said. "But they want to go with us. Do you think that is a good idea?

"They're desperate and need help. We shouldn't turn away hungry people when they are right there in front of us. We'll take them with us. How many of them are there? Fifteen or so? We have room. They can come and live with us. Besides, at some time in the future, we may want to establish relationships with the people along this coastline. They may be helpful in doing that."

* * *

They had timed their flight that day to arrive at night and land in the gardens, where they had landed on their first trip. Landing at night ensured that fewer people would see them.

"I pulled the geo-log from the shuttle from our previous flight to GK's kingdom, so I know where to land. Also, we are deliberately landing at night to avoid being seen by masses of people," Hatcher explained to Connie, who had asked about how the people would respond to a flying machine landing in their capital city.

When they descended the ramp to the ground, they were greeted by a group of armed guards. They didn't see him at first, but GK was back behind the guards. He walked and greeted Hatcher and Gunner, whom he remembered.

"Welcome, my friends! We are so happy to see you again, though I must say, these are not the happiest times. The climate is changing, and this means the weather is cooling. Our crop production is not what it used to be. I'm afraid we may go hungry. Now, who is this here, this lovely lady?" GK asked when he saw Connie standing behind Gunner and Hatcher.

"Oh, yes, GK, this is my daughter, Connie. She wanted to come to see your kingdom. Connie, this is Genghis Khan, or GK, the emperor. He got to know your mother from our visit here some years back."

"I am pleased to make your acquaintance, Emperor." She held out her hand to shake his extended hand, but instead, he took her hand in both of his and kissed her on the hand, mesmerized by her beauty.

"Please, Gunner, Hatcher, and my dear Connie, come with me to the palace. The remainder of your people may have to wait until tomorrow for me to arrange accommodations for them. You have come unexpectedly, I'm afraid."

"That will be acceptable, GK. Lead the way," Gunner said.

When they reached the palace, GK led them to the same rooms they were in before. They waited for morning to discuss their purpose with him.

* * *

Connie woke up in the morning to find two servants in her room, waiting for her to wake up.

"Hello, I am Connie. What are your names? Why are you here?"

"Princess," one of the servants said, then paused and continued. "Princess, we are here to help bathe you and dress you. We have appropriate clothes for a princess. When you are dressed, we will take you to breakfast with the emperor and your colleagues."

"Oh, I can do all that, I'm fine …" but she was cut off.

"The emperor insists. Please, here is your bath. Let us help you with your clothes."

"All right. Umm …" Connie stood, and then the two servant girls began pulling her pajamas off, leaving her naked and feeling vulnerable. She decided that if she went into the tub, she would be less exposed, so she stepped into the tub and lay down.

They began washing her body with the same lavender fluid that Terra had been washed with some years ago. Connie decided she really liked the smell, and it made her feel so *relaxed.*

"I'm not sure what this purple liquid is, but I love the smell! And I feel so strangely calm," Connie said.

After finishing, they dressed her in a beautiful blue and white full-length gown, which had a somewhat low-cut front. They combed her hair and tied it up in a way she had never done. Connie felt somewhat uncomfortable with the low-cut dress. She had never been one to advertise her figure. She was barely eighteen now, which was considered an adult, though she still felt as if she were a child in many ways. She worried if her father would approve of her hair and dress.

The two girls followed her down the wide, winding stone stairs to the dining room, one floor below from where she had slept. At the long table, the emperor sat with Gunner and Hatcher who were seated with him. All three stood as she descended the stairs. Connie didn't know it, but Gunner and Hatcher both noticed how grown-up and beautiful she looked. She no longer looked like the adorable little girl they once knew. GK was clearly smitten with her beauty as he could not take his eyes off her.

"Sweetie, you look so beautiful! You look so much like your mother. She descended those stairs a few years ago and looked beautiful, just like you!" Hatcher said, beaming with pride at the sight of his little girl, who was no longer a little girl.

"Please join us down here, Connie. Is Connie shortened from something longer?" GK asked.

"My full first name is Constance, my emperor."

"Please call me GK, dearest and most beautiful lady."

"So I shall, my emperor, GK!" Connie said, laughing and smiling, accentuating her freckles, the only thing she inherited from her father.

As they ate, GK asked about their purpose in coming to his kingdom.

Gunner answered. "GK, we have noticed that the climate has changed as well. This has also made our crops grow poorly. We brought a great many things to trade, hopefully, with you. We are seeking fruits and vegetables. We have salted salmon, a fish that is delicious and plentiful in our waters. We also have salted turkey, a tasty fowl that we believe you will enjoy. Also, we would like to share some ideas with you. The Terrans we brought with us are noted shipbuilders. Their ships are able to ply the great waters of the oceans, not just rivers and lakes. These ships could help you trade goods with us in the future. They would like to build a ship for you and train your people to build them on their own in the future. Also, we have radios we could leave here that would allow you to communicate with us should you ever have an urgent need, such as a medical emergency or if an enemy threatened you."

GK nodded his head and leaned back in his chair. "I can see the merit in establishing trade with your people. We could spare some of our fruits, such as oranges, nectarines, and grapefruit. We have vegetables as well—turnips, beets, squash, watermelon, and carrots. Our people are in need of meat. I believe you say meat is rich in protein, and human beings need a portion of their diet to be meat or fish. I think we can reach an agreement on this. My trade minister will look at what you have and come up with a fair exchange. I am very interested in this thing you call a 'radio.' I could have my wise men look over your devices. It would be most helpful to learn about this device from the basic principles. Gunner, who could help my people with this?"

"I know Connie is studying the field of engineering. She could instruct your people so much in this area. We do have some limitations put on us by our great Mother, Terra. She wants to slowly introduce technology to the Terran peoples. We do not want to introduce weapons into the Terran groups. It is our desire, above all else, to establish peace between all the Terran peoples. We know that sometimes it is necessary to use force to repel an enemy, but that is something we would help you with only if it were truly a great need. Radio communication, however, would mostly be employed in peaceful endeavors such as aiding in the integration of your empire and our colonies."

Gunner and Hatcher stood and shook hands with GK. Connie, who was also at the table with them, shook his hand as well.

"Connie, it would please me greatly if you could remain here in my capital after your people leave. I'm sure it would take quite some time to teach us all the things you know. Would you be willing to do this?" GK asked.

"I will speak with my mother and see if she agrees. I would like to stay. There are so many interesting things here to see. I would especially like to read about the early history of The Mothers here on Terra Nova, my mother told me she found it fascinating. She would like to return one day and read more in your great library."

"Perhaps you could stay here for several moons, and we could return together on your great flying ship. I would like to visit your colonies and see your mother again. I grew quite fond of her, though I see so much of her in you."

CHAPTER 51

Mission Accomplished

"MOTHER SAID I could stay. She thought it was a good idea. I'll be fine. I can learn so much here, and I can help them set up a more powerful transmitter. I'll be busy helping them build a water wheel to spin the turbine you are leaving me with. This is going to be fun!"

"I know. I'm just nervous about leaving you here almost alone. I know Jenny will be here with you. Maybe I shouldn't worry so much. Call on the radio every couple of days to let us know how you are doing. The battery will power the radio for months, maybe a year or more," Hatcher said.

"Well, I plan to get a source of electricity up and running soon. You left me with enough wire to run power lines to the palace. I've got excess LED bulbs, so I can start to electrify this place. There are so many technologies involved in creating a modern society. I mean, making LED bulbs is quite a task. I've got all the AI technology files. I can be the Thomas Edison of Terra Nova!"

Hatcher smiled at his daughter's enthusiasm. She saw only adventure and fun. Hatcher saw other issues she had to navigate. He knew GK was sorely smitten by Connie. Hatcher was worried about this. She was only eighteen years old—too young to get married, or so he thought.

How would she handle the attention of a powerful man like GK? Would she make a wise decision? I must remember she is like her mother. She has "the gifts." She was learning to use them. There was so much that could be done with the people here. They don't have modern technology, and I know we have to introduce it slowly so that they can learn to use that technology wisely. Well, that is our job here, it seems. We are supposed to help develop them into an intelligent and virtuous species. They certainly seemed to be capable of intelligent thinking, Hatcher thought.

In the end, parting was painful. Hatcher was losing his little girl. He knew it wasn't forever, but distances were more intimidating on a primitive world like this. To see her meant coming in the shuttle. Shuttle missions were tightly monitored to conserve on fuel and wear and tear on these complicated machines. He wasn't sure how long it would be before he saw her again. GK did say he wanted to visit Benston.

How would he come? On one of the ships, the Terrans from Ammonsberg had built for them? Some of these builders were staying behind to continue helping the people here learn how to navigate on open water. I am worried about her traveling across the great Terran Gulf in a small boat. We're leaving after staying for almost fifty days. We've emptied the shuttle of the dried salmon and turkey, and many other technology items that we traded for fruits and vegetables. This mission was a success. We're also returning home with some new people we picked up on our way here. All in all, we had a successful trip. Mission accomplished. Hatcher thought.

CHAPTER 52

Rationing

THERE WAS SPECULATION that somehow the climate might be heading into an ice age. Why that was the case was the subject of some debate among the scientists who had expertise in climate systems. All that mattered was that the weather had not yet returned to what it had been like leading the scientists to believe the climate might be changing as well.

The shuttles had made several trips into orbit to analyze the glaciers and polar ice coverage. The conclusion was that the glaciers had grown, and the polar ice at both poles had done the same. This was some confirmation that the planet may be entering a glacial period with a glacial maximum perhaps hundreds of years hence. No one found that comforting. They lived at a temperate latitude. Some wondered if they should move toward the equator, but land in those latitudes was still a jungle, and that brought more problems than solutions. They would stay put for now.

The colonists had switched crops to ones that grew better in cooler conditions. These included cabbage, peas, and certain strains of potatoes. Additionally, they were building more greenhouses to grow crops over the winter to make sure they could last through another lean season.

At the next meeting of the supply council, the status of food stores was discussed in great detail.

"By the time next spring comes, we should plan another trip to GK's kingdom. We'll have salted away more than six tons of salmon and three hundred turkeys. That will be worth quite a bit in terms of fruits and vegetables by trading them. Even so, we are going to have to carefully ration the food we have. We have over 50,000 people here now, and with 1,800 calories per adult and 1,500 for children under twelve years of age, we will just make it to spring. At that point, well, I'm not sure what to do. Maybe the greenhouses will produce enough for two to three months, but at that point, our reserves will be exhausted," President Stanton observed, referring to the report produced by the supply council.

"We may need to send shuttles on supply runs to the large southern continent. We believe a great many tropical fruits grow naturally there. To harvest enough for our needs is going to require a couple of hundred workers working for a month or more," Peter noted.

"We'll need to issue ration cards for everyone. For children, the parents will be responsible for holding their children's cards. I know none of us like this idea, but I don't see any better way to ensure the food we have is equitably and fairly divided up among everyone," Carla Chan said, who was the chairperson of the supply council.

“What about our investigation into introducing more greenhouse gases into the atmosphere to raise temperatures?” President Stanton asked.

“It would work, but it may take several decades before we see any increase that would make a difference,” Dr. Kimball said.

President Stanton listened intently, then said, “So in the near term, we are just going to have to find other food sources here and elsewhere.”

“I’m afraid so, Mr. President,” Terra said.

CHAPTER 53

Electrification

THE PROBLEM OF creating a power grid wasn't that the principles of electric generation had not been discovered. The problem was that raw materials and production processes were not yet developed, Connie thought. *I'm going to have to find deposits of copper and figure out how to fabricate them into wire. Add to that the need for steel. Iron was probably plentiful, but where do I come up with carbon to add to the molten iron to make steel? Aluminum is even more difficult to make. The power lines will be made of steel for strength and then aluminum cores to carry the electricity. I have never even seen any of these processes. Well, I do have the AI to lean on, but it is going to take labor, and the labor will have to be trained in what we are doing. It makes sense to train them along with me. I'll need to ask GK for people who can learn these complex things.*

"GK, to create a power grid is going to require quite a bit of labor. We'll have to create new industries, ones that don't exist now. We need to have access to deposits of copper, and then we will have to draw the copper into wire, in very small diameters. We'll also have to develop a means of drawing flexible plastic around the drawn wire. Wow, it is a bit bigger job than I imagined," Connie explained.

"What is plastic?" GK asked.

"It is a material made from petrochemicals. I know you have tar pits. I think we might boil water, and that will allow the bitumen to be separated and float to the top of the mixture. This can be fabricated into plastic. We also have to develop the means of extruding plastic. Plastic is needed because it does not allow electricity to flow or pass through it. It insulates the conductor, which is made of copper. The plastic allows the copper wire to be handled safely when it is conducting electricity. It does this by blocking electrical charge from leaving the copper wire. It is like a vase that allows water to be carried, without leaking and letting water spill out."

"You're really enthusiastic about this, aren't you?" GK noted, smiling at her.

"Well, having an electrical grid is the basis for so many other advancements, so yes, I want your people and your kingdom to lead the way here on Terra Nova."

"I can see you care a great deal about making this happen. Let me know how it is going every so often. If you need more help, I can always get more people to help you."

"Thank you, GK."

* * *

"Daddy, I need to talk to Peter Chan," Connie said over the radio to Hatcher.

"I can find him and have him call you in, say, three hours from now. Will that work? What do you need to talk to him about?" Hatcher asked.

"Making copper wire. This process is going to be crucial for us in the long term, and it is necessary for making copper wire for the electrification of the larger cities here on Terra Nova," Connie answered.

"So how is the project going?" Hatcher asked.

"Well, I have demonstrated we can produce electricity from the river that runs through the city. The water wheel turns the blades of the turbine. Steam is also directed at the blades to cause them to rotate. The magnets on the shaft rotate and induce current in the wire coils. It lit an LED bulb, but that is a far cry and a very small step in the huge process needed to produce the electricity we would need for this entire city here."

"I see. So you want to make large quantities of copper wire. I suppose you'll want to know how to extrude PVC onto the wires as well?" Hatcher asked.

"Yes, but one thing at a time," she said.

"This is going to be a long-term project. We may want to have each Terran group specialize in certain parts of this production process. We can make the tooling needed for making copper plates, then copper wire, and so on. We have power here and machine tooling like lathes. I can see that if we get this industry up and running it will provide the power needed for other advancements. Does your mother think this is too much technology too soon?" Hatcher asked.

"Could you ask her about that, Daddy?" she asked.

"Sure, will do. I'm proud of what you are doing, sweetie."

CHAPTER 54

Assassination!

RAPHAEL WAS GIVEN the change by The Mothers after he arrived on this world thousands of years earlier. With the passage of time, he began to believe his talents were not being fully utilized. He made friends among other males that had come with The Mothers, and some of them he associated with stroked his ambitious nature, which he had not yet tamed, even given the centuries that he lived on Terra Nova.

He watched Connie and GK closely. He could see what they were doing, advancing the technology of GK's empire. He believed that he could still count on a rescue if only he had a radio. He knew he needed to find the stash of high-tech items that his superior, Michael, had hidden away in a cave somewhere. He knew it included an EPC device. If only he could find it. He was still looking, but every cave he looked in was a dry hole, so to speak.

He had worked his way onto the mining detail—the group of workers who were mining ore that contained copper. The group met weekly with Connie, who reviewed their progress. As soon as they had enough raw ore, they were going to begin working on the process of separating the copper from the ore. He knew how to do this. He had been trained as a mining engineer on the AOW's world before he was dropped off on Terra Nova with the other two thousand AOW spies who blended into the population of the Empire of the Stars. By the time he got there, that empire and all The Mother's people were being withdrawn, as the work from that point on was put into developing the final genomic changes to the hominids on Terra Nova. That final work was to be completed later, when The mother of the new world would arrive. No one knew when that was, but the rumor was that it would be someone from an advanced world, but not from The Mother's world. This was a bit of a twist on the normal pattern followed. This new individual would be performing their first-ever tasks in the name of The Mothers. He now knew who that individual was—it was a young woman named Terra who came from the world called Earth. She had come on an odd ship—an asteroid built into a starship. Their technology was not up to the level he knew The Mother's world possessed, but it was still quite advanced.

The work he and his colleagues had accomplished was impressive. The time between when The Mother's people were withdrawn and when the Earth people had arrived, had been put to good use. They had successfully given gun technology to the people of the plains. But alas, he was the only one who was given the change and he was now alone. He had been "volun-told" that he must remain behind to watch for the moment to kill the new Mother and any of her offspring. He now had a golden opportunity to take out the daughter of the new Mother. He strongly suspected Connie possessed the gifts as well, having inherited them from her mother, Terra. If his plan worked out, he could take them both out at the same time. That time was drawing near.

CHAPTER 55

The Bomb

RAPHAEL HAD HEARD that the Earth people would be returning. The work on building an electrical grid was, of course, nowhere near complete. The Earth people wanted to trade regularly with GK's kingdom. They needed fruits and vegetables, and GK's people needed protein from the salted meats that the Benston Colony had aplenty. He also learned that their chief engineer, a man named Peter Chan, would be coming. Apparently, Benston was building tooling for the fabrication of wiring from copper and insulation for the wiring. Honestly, Raphael was impressed with the Earth people's industriousness in trying to establish an electrical grid from scratch. It was one thing to know the scientific principles, but it was an altogether much more difficult thing to build the industries needed for the components. The AOW spies had never even attempted such an ambitious endeavor.

Their arrival was three weeks away. He needed to finalize his plans for the assassinations. If possible, he wanted to take out as many of the Earth people as he could, along with the two Mothers. He planned to kill them all with a bomb, a big bomb. He knew there was going to be a big formal dinner the evening after their arrival. That would be his chance. He had already volunteered for the serving staff.

He had spent months getting the mineral gwihabaite from the deposits found in the empire of the horsemen. From this natural source of ammonium nitrate, he would build his bomb. Fuel oil he obtained from the tar pits that the electrical grid project was using for making insulation for the wiring. Detonators would have been a much more difficult problem, but the AOW had a weapons stash that still had some of these small devices. They were now very old, but they had been kept cool and dry, so he hoped they still worked. He had tested several, and most of them worked. He planned to employ complete redundancy and use two detonators. The bomb would be a true detonation, not a conflagration. The detonation would produce a shock wave, which would destroy the lungs of those within two to three hundred meters. The fire would finish off any survivors. He needed a timer, and fortunately, the AOW stash also had a few of these. He wanted to be well away from the dining hall when the device went off. He even tested a subscale bomb several months earlier. It worked exactly as he hoped. It seemed he had everything he needed.

He planned to put the bomb in a crate that held the dried salmon and turkeys the Earth people had used in their first trading trip. It would be put in the kitchen store room for the big formal dinner. At the appropriate time, he would activate the timer and make his getaway.

CHAPTER 56

Dinner, an Announcement, and Death

CONNIE WAS NERVOUS. She had not told her parents of her big news. She planned to tell them at dinner that night but wasn't sure how they would receive the news. She had now been in GK's kingdom for almost a year. This latest trading mission was the first time in a year that she would see her parents. She missed them, and she knew they missed her. Her big news meant she would only be coming home for short visits in the future. GK's kingdom would now be her home. She and GK were getting married a year from now, which meant her parents would have to make another trip here to see that. She hoped that would coincide with another trade mission. She didn't want to make them come out just for the wedding. She was like her mother in that she didn't want to impose on other people or have special privileges, though when she became the wife of the emperor, she would experience constant special privileges!

On the night of the shuttle's arrival, Connie spoke with her mother in the gardens where the shuttle had landed again.

"Connie, it is so good to see you! How have you been?" Terra asked her daughter.

"I'm good. I see you have another little one on the way! Boy or girl?" Connie asked.

"Twins, again! Another boy and a girl. Willie stayed behind to watch little Timmy. He really wanted to see you, but I just didn't feel good about bringing Timmy on a long trip; he gets into so much trouble now! Taylor is also expecting, literally right now. Your father needed to stay with her, of course. So Willie was the logical person to watch Timmy."

"Well, I guess he is in the terrible twos or threes or whatever they call it!" Connie said.

"Yes, for sure. Although he is not nearly as precocious as you and Willie were. You two were into everything. I had to watch you two like a hawk!

"Well, I am so tired. I think I will go to bed early tonight. The trip here was so long, twelve hours. We'll talk more in the morning, sweetie," Terra said.

"Good night, Mother."

* * *

The next morning, servants began preparing the great dining hall for the evening events. Raphael had several men, seven in total, who were helping him. They were men who had a grievance with GK. They also had a plan for after GK was "removed." There would be political turmoil and uncertainty. There was no clear-cut successor. The old regent who had run the kingdom while GK was too young was now quite old. One of the men who worked with Raphael was a man named Ishmael, who was a cousin of GK. He felt he had the best claim as a successor. If he succeeded, then each of the others were guaranteed high positions within the new power structure. There was more than enough motivation to keep them all involved.

The box with the bomb was moved into the storeroom of the kitchen, which was only ten meters from the dining area. The distance ensured that all those in the dining area would be killed. In fact, everyone in the palace would either be killed or gravely wounded. The plot was sure to succeed if the bomb exploded at the right time. The right time was just after the main course was served. Probably right after GK made his big announcement, which was hardly a secret to anyone who worked closely with GK.

At seven bells past noon, the guests began arriving in the dining hall. Tables with alcoholic beverages had been set up so that those who came early could properly socialize, and many of them were early and began drinking. By eight bells, the guests were seated. An appetizer of peaches and vegetables was served. A group of musicians softly played harps and flutes to fill in the background din.

In the kitchen, Raphael and three of his men excused themselves to go to the storeroom where they had secreted the crate with the bomb. They were going to put the detonator into the fuel oil and ammonium nitrate slurry and set the timer for ten minutes. That left them just enough time to leave the palace and reach a safe distance. After the detonator was set into the explosive mix, Raphael set the timer, and the three men told the head of the serving crew they were going to take a break on the balcony. They did not head to the balcony. Instead, they left through a back entrance to the kitchen and headed through the palace grounds after ditching their serving clothes. If they were lucky, no one who saw them would be alive to say that they saw them leave. This would have worked except that Terra had excused herself to get some cool night air. She was nauseous and didn't want to throw up on the table and food. She was a little over twelve weeks pregnant again.

GK stood at a quarter past eight bells and welcomed the guests and stated he had an announcement. He was seated at the head of the large, long table. Connie sat next to him, dressed in a beautiful gown with flower prints. Her hair was adorned with flowers, and she wore a beautiful necklace that GK had given her as a birthday gift just a few days earlier.

GK, who had not noticed that Terra had quietly slipped out, stood. "I want to welcome you all here this evening and thank you for coming. I want to especially thank our guests who traveled a great distance from the Benston Colony to be here. I have a happy announcement to make. Many of you have noticed that I have been preoccupied during this past year. There is a very good reason for that. My dearest Constance and I will be married before the end of next year!" The guests began to clap as GK raised his cup of wine and proposed a toast to his bride-to-be. The guests raised their cups as well.

At that very moment, Terra sat down on a bench in the courtyard. As she sat, she saw several men leave through the back door of the kitchen. Terra found it odd that anyone would be leaving as the main course of dinner was yet to be served.

Terra froze as images came into her mind of the thoughts of some of the men. She saw them thinking of what they had just done—arming a bomb! Terra stood and ran into the dining hall.

As she ran in, Connie caught sight of her mother and instantly saw the thoughts of her mother. She saw great danger in her mother's mind, and a single word caught her attention: *bomb!*

Terra began to shout. "Everyone, leave the dining hall now! There is great danger. A large bomb is about to go off. Leave now! Run!"

GK, who had just sat back down, stood up again. He was confused. He did not know what a "bomb" was. Others began standing as well, sensing that something bad was about to happen. The Earthers, who knew the meaning of what a bomb was, began shouting for everyone to leave.

There were two entrances to the dining hall, and they soon became clogged as more than 250 guests tried to leave all at once. Connie tried to tell everyone to move quickly but orderly, but the obvious distress of Terra was not lost on those who knew that Terra and Connie had special abilities, and they were clearly frightened, which caused those in the hall to become frightened as well, even though the Terrans did not understand the exact nature of the danger.

When the guests reached the outside courtyard, they continued running to the nearest palace gate. Guards at the gates were confused and tried to stop some of the guests to ask what was happening.

Terra had grabbed her daughter's hand as they left the dining hall. Connie had grasped GK's hand as he left the hall with her. The three of them were running for the gate when the bomb went off.

The first thing anyone noticed was a loud pop that burst their eardrums, followed by a flash of light, then nothing. The shock wave enveloped the fleeing guests and burst the lungs of those closest to the dining hall. They died instantly. The others sustained very serious injuries as the shock wave filled the air-filled spaces inside their bodies and compressed their organs, leading to many dying a few minutes later. Through sheer chance, some were not as exposed to this deadly effect, but their injuries were nonetheless grave.

Several minutes went by before any of the injured regained consciousness. Silence ruled in the aftermath. Eventually, some of the injured regained consciousness. No one could hear properly. All they remembered was silence or a loud ringing in their ears. Those who could stand did so, only to fall back to the ground as their sense of balance was destroyed by the injuries to their vestibular system.

Eventually, guards and other palace workers who were lighting evening torches on the grounds came into the courtyard and began trying to help the injured.

Connie, who had been one of the lucky ones, had a concussion but no other injuries, including burst eardrums. Terra was in a similar condition. They began looking for GK, who they found lying on the ground face down. He had not been so lucky. He had internal organ damage and was still unconscious. Connie had several of the guards who ran into the courtyard turn around and run to the shuttle in the gardens to get the two doctors who had come on the trip and others who could help with the injured.

The scene was one of total confusion, with hundreds injured and more than a hundred killed instantly. Terra realized they were going to have to get the most gravely injured to the shuttle and take them back to Benston ASAP to have any hope of them surviving. She sent another two guards to the shuttle to tell them to prepare to leave on an expedited flight path. The race against time had just begun.

CHAPTER 57

The Race Against Time

CONNIE, TERRA, AND a number of the medically trained individuals were on board to help the injured as they were loaded into the lower deck.

Dr. Sampson was again in the middle of a tragedy. As the injured began arriving in the shuttle, it became clear to her that there was, again, little that she could do. The injured needed the facilities that only Benston had. She did what she could, but when she found Terra and Connie, she told them that she thought her best course of action was to stay here and help the less seriously injured. In fact, there were four other nurses who volunteered to stay and help take care of the less seriously injured with Dr. Sampson. All told, there were ten Earthers who thought they could help the situation by staying and returning later.

At the moment, there was no one in charge. The old regent thought it prudent that the eldest member of the emperor's staff be put in charge temporarily. This man's name was Quiang Chen, the finance minister for GK. He assumed the duties of the emperor, and his first task was the treatment of the wounded. His second task was to apprehend the guilty parties and administer justice to them. He felt that this was what the emperor would do, and he began to pursue these courses of action. Terra had told them about the vision she had from the minds of the men who were leaving the kitchen area just before the blast. The kitchen staff was held in custody, and it was determined that several were missing. Minister Chen assumed these were the guilty parties, and he authorized the military and palace guards to begin looking for them. At the time the shuttle departed with the most seriously wounded they had not been apprehended.

At 5:00 a.m., the shuttle blasted off from the palace gardens in an expedited return flight path. Again, the liquid fuel put the shuttle on a suborbital ballistic return trajectory that would take one hundred minutes to reach Benston. The remaining doctor on board was torn between returning quickly and risking further injury from the g-forces or returning more slowly over a twelve-hour period. In the end, she decided that they would lose a roughly equal number of injured regardless of the method used to get the injured to Benston.

From the flight deck, Gunner and Peter manned the controls and were again going to manually fly the shuttle "up and down the line" following the acceleration curve projected in front of them via holographic projection.

Gunner spoke through the intercom. "Five, four, three, two, ignition sequence started, one, liftoff!"

Everyone on board, including the badly injured, pulled several g's as the shuttle engines ignited and the shuttle accelerated quickly into the early dawn sky. For four minutes, the engines burned, then cut out. The shuttle reentered complete darkness as it rose above the atmosphere and began arcing over the horizon. Ninety minutes later, the shuttle reentered the atmosphere and began glowing brilliantly as the heat built up on the exterior of the craft from friction with the atmosphere. The final four minutes were filled with deceleration forces as the shuttle engines burned furiously, slowing the craft to a safe landing speed.

Soon, the door ramp was lowered, and within a couple of minutes, the injured began being removed by waiting medical personnel. Their fates were now in the hands of the doctors and the Almighty.

CHAPTER 58

Meanwhile, Back at the Ranch

DR. LYNN SAMPSON was beyond worn out, and so were the nurses who stayed behind. For four straight days, they had been treating "the less severely injured." Lynn laughed to herself at that description. A full one-half of the patients they were treating were now deceased. Lynn had never been in a war scenario as a physician. She could not imagine how it could be anything different from what she had now experienced.

She did not have the benefit of much in the way of modern medicine and equipment in this situation. Lynn enjoyed genealogy. A distant ancestor of hers had been a young surgeon in the first American Civil War, which, until the second American Civil War, had been America's bloodiest conflict. The second civil war far exceeded that. The early twenty-first-century president, commonly referred to as "the traitor," had been, according to most historians, the primary cause and architect of that war, even though he died in prison before that war began.

Why exactly was he in prison? she thought. *Oh yes, he was impeached and removed from office for pedophilia. What a total creep,* she thought. *That was what finally broke the cultish hold he had on his political party. Even they couldn't look the other way on raping children! Those were incredibly perverted and tough times in American Middle Times history. Well, we got a new constitution in the aftermath of that war, and it mandated unbiased psych evals on all candidates for federal office as well as deep background checks. Shit, that guy had sold the codes and deployment locations for our ballistic missile subs to the Russian leader! That led to the second near-miss of total nuclear war in 2027. Those Russian fuckers actually thought they could win a nuclear war with us because they knew the codes and locations of our stealthy nuclear subs! The whole thing unraveled badly for the Russians and the American president when he realized he had been given dummy codes and location data in a deliberate deception authorized by the Joint Chiefs of the US military, because they were worried explicitly about this kind of treachery by this incredibly flawed man! Not all our people's programming had been defective. Thank God for true patriots who loved their country more than protocols, laws, and that flawed old document, the first US Constitution! History, she remembered, put this incident in the same category as the Cuban Missile Crisis of 1962. Did The Mothers or The Ones somehow intervene to save us then? I need to ask Terra about that someday. Well, the Russians eventually got what they had coming when their own chief ally, China, a century later, performed a surprise nuclear attack on Russia that resulted in both countries being destroyed, well, and the Earth in the end, sadly.*

That night, none of them could go any farther, and they all fell asleep in one of the palace's guest rooms. Many of the palace staff were still alive, though some, sadly, had been among the injured and had died.

Minister Chen had thus far been unsuccessful in locating any of the missing kitchen workers. They seemed to have blended in with the local population.

* * *

At one bell, Dr. Sampson was awakened by a palace guard and asked to hurriedly come to the emperor's room. Minister Chen had been sleeping there since the day after the bomb went off.

"Please, we need you to examine the minister! He will not wake up!" the palace servant explained once she got there.

Dr. Sampson examined the minister, listening to his heart and lungs. His heart rate was very slow, and his breathing was also very shallow. He appeared to be drugged. She could not rouse him by shaking him. She administered a stimulant, which she hoped would increase his alertness as well as speed up his metabolism. This had no effect. She drew blood and ran it through her mobile pathology unit to see if she could identify the agent that produced this condition in the minister. Lynn strongly suspected that he had been drugged intentionally. There was simply no other reason that could explain his lethargic state. The results took several minutes. The result was a cardiac glycoside. She knew there were several naturally occurring plants that contained these substances, such as foxglove, lily of the valley, and oleander bushes. She had seen oleanders in many locations, including on the palace grounds. Unfortunately, she did not possess any antidotes or the means to produce one. All she could administer, from what she had, was activated charcoal, but given the depth of his physical state, she knew it would likely not be enough to rouse him. He was going to die, and there wasn't much she could do about it.

"Can you tell me if there are any other cabinet ministers left alive?" she asked the servant.

"My dearest lady, Minister Chen is the only surviving minister," the servant explained.

"Minister Chen will very likely die, probably within the next two hours. I am not sure if you have someone else who can assume the duties of GK, but you should try to come up with someone who can perform the duties of the emperor. Is there anyone? Perhaps a relative of GK?" she asked.

"The emperor's closest relative is his cousin, Ishmael. I can send someone to summon him. He lives nearby," the servant explained.

"Please do so. It is most urgent."

Within the hour, Ishmael was brought into the emperor's room, where Minister Chen lay dying.

"Ishmael, I am sorry to trouble you at this late hour. Minister Chen has been poisoned. He will die, very likely, within the hour. I do not have the means to help him here. If he were in our medical facility at our colony, I could easily treat him. It is not possible here. I am told that no other ministers survived the explosion. It is not my place to tell you who should assume the duties of the emperor, but perhaps you are familiar with the rules of succession. Someone must be selected soon. Your kingdom has suffered a devastating terrorist attack. There must be leadership in place to deal with the many pressing issues that need attention. What do you suggest?" Lynn asked.

"The rules of succession dictate that the closest adult male relative of the emperor assume the mantle of emperor. I am, unfortunately, the closest adult male relative of GK. I must reluctantly and sadly assume these duties. I appreciate all you have done for Minister Chen. I must summon a judge to witness these events and validate that I am the rightful successor. I will need witnesses to attest to what has happened. I will need you to attest to the cause of death of Minister Chen. Will you be willing to do so?"

"Yes, Ishmael, I will explain to the judicial authorities what has happened."

CHAPTER 59

Fighting for Life and Kingdom

GK HAD JUST undergone four hours of surgery to remove his spleen and stop internal bleeding. His internal organs had been damaged, and no suitable treatment for replacing his damaged liver, lungs, and kidneys would help him in the short term. Given time, new organs could be grown for him, but that would take quite a while. He would have to fight for his survival on his own. His liver was still functioning, marginally. His left lung worked well enough that it continued to function, though he had been intubated and put into an induced coma due to the extreme pain and discomfort of his injuries. It would be touch-and-go for him for days, perhaps weeks. No one was giving odds on his survival. He was getting the best care that was available on this planet.

"Dr. Le Cesne, he seems to be doing better. His liver function has increased somewhat. Is this a good sign?" Connie asked.

"Yes, it is, my dear, but his kidneys are what concern me the most. Right now, we have him on dialysis. When, and if, his condition stabilizes and he improves enough, we can extract stem cells and begin to grow new kidneys, a new right lung, and a liver. We need to nurse his liver along to keep it functioning so it can do its job while we grow a new liver. That, as you know, will take upwards of three to four months, so he needs to keep going that long so we can grow him new equipment!" Dr. Le Cesne said with a smile on his face. "I am more positive now than I was a week ago. Let's keep thinking positive thoughts, shall we?"

"For sure. He is the love of my life. I don't know what I would do without him!" Connie said as she began weeping again.

"Connie, I can stay with him tonight at his bedside if you want to go home and get better sleep," Terra said.

"Thank you, Mother. I do need sleep."

* * *

"What did Lynn say when she radioed in last night?" Hatcher asked.

"She said that Minister Chen was poisoned and he died. She didn't have the necessary treatments available to her to save him. She also said the cousin of GK, Ishmael, who is now the emperor, is a real snake in the grass. She doesn't have proof, but she suspects he killed the minister. She wonders if he had something to do with the bomb. You should know she is worried about her safety. The people who have survived the bomb blast are going to recover. Those who died, well, they were beyond saving with what she had available to her. I think we need to go get our people. We should talk to Terra," President Stanton said.

"Well, if it were up to me, I would go get them right now. Somebody employed brutal means to try to kill GK. We lost six people, and a dozen were badly injured. We are leaving them in a bad place," Captain Simmons said to President Stanton. Captain Simmons was an advisor to President Stanton.

"All right, let's go get them. Sandra, can you and Gunner make a plan to go retrieve our people while accounting for the possibility that we may need to use force? Hatcher, I would like you and me to go talk to Terra. Let's see what she thinks. I know she is always reluctant to use force, but we may end up in a situation where we have to do that."

"OK, when do you want to talk to her? I know she is planning to stay at GK's bedside this evening. She wanted Connie to get some good sleep at home," Hatcher said.

"Let's do it right now. I think time is of the essence."

* * *

As they entered GK's hospital room in the Benston medical facility, they were surprised to see GK awake, without a breathing tube. In fact, he was talking to Terra and Connie.

"Well, well, look who has finally woke up!" President Stanton said lightheartedly."

"Yes, it is good to be back and awake again. It is hard to do things. I seem to be out of breath all the time."

"Well, sweetheart, you have only one good lung. But that will get fixed up eventually!" Connie said.

"GK, I guess you can listen in on what we are going to talk about. We wanted to talk to Terra about going to get our people. Apparently, Minister Chen was murdered. He was poisoned. One of your relatives has assumed the role of emperor ..."

GK interrupted President Stanton. "Ishmael. That is not good. I fear he may have had something to do with the bomb blast, and yes, I believe—but do not have evidence—that he murdered Minister Chen. He is a man of low character and has been jealous of me my whole life. He constantly picks at me with little comments, criticizing everything I do. I have spoken to him privately, but he always denies that he is trying to undermine me, but he is. Your people could become pawns in his plans to hold on to my kingdom. You do not want to let him know you are coming. He may hold your people against their will to gain leverage over your people here. I must get well soon and confront him. I—" At this point, GK slumped back into the bed, in obvious pain.

"GK, my dear, do not trouble yourself with these things. We will get our people, and at the right time, when you are well, we will help you regain your kingdom, my love," Connie said.

"Our people are in danger. They are already being held against their will. We must act now to rescue them. He will resist this and will only give up our people through force. The people in GK's kingdom love him. They fear this new ruler. If it becomes necessary, he may need to die for his actions to prevent and deter further bloodshed. I do not believe you can rely on information from our people on the radio. They will be forced to lie to us," Terra explained somberly.

* * *

"Terra says that our people are in a palace guest room on the upper floor of the palace. There are guards leading up to the floor and outside the door. Guards are also stationed outside of the palace structure on the ground level. Terra says they want Connie to be given to Ishmael in order to get our people back. This all sounds so familiar," Hatcher lamented.

"What can we do to get them out without causing a bloodbath?" President Stanton asked.

"Mr. President, I am told that GK is loved by his people. If GK were able to go along with us, he could speak to the palace guards and other staff. Perhaps that would convince them that they should abandon this imposter in favor of returning GK to the throne," Captain Simmons said.

"Everyone, GK is still not well enough to travel. I could go and speak to the palace guards and staff. I can push thoughts into their heads that GK is the rightful ruler and that supporting the imposter will lead to the fall of the kingdom and the loss of their freedoms, personal prosperity, and well-being," Connie said.

"I am well enough to go, though do not expect me to be jogging around the palace grounds! I know Ishmael. I can convince the people, the palace staff, and others that I am alive and ready to resume my duties, though that will be a ruse. I will have to appoint a regent to run the kingdom as I will need to return here and finish the medical treatments that the doctors have in mind for me. First and foremost is to get a new second lung!" GK said as everyone stared at him, wondering how he had managed to stand up and walk into the room where they had met.

"You must forthwith return to your bed, GK! I will not have you pushing yourself like this," Connie said insistently.

"I shall do as you wish, but I do not want all of you to be fighting my battles for me. I will do what I can, and what I can do is to reason with Ishmael. I believe someone else has emboldened him to pursue this reckless course of action. I have been told there was a man, a member of the AOW, who is immortal and resides in my kingdom. The rumors are that he was left here, on this world. He wants badly to return to his world but lacks the ability to communicate with his people. If this is true, he could be the one who precipitated these calamitous events, the bomb, and the poisoning of Minister Chen. The AOW people were almost single-minded in their efforts to bring down the work of The Mothers—rumor has it, anyway. We should all hope he does not somehow get the capability to communicate with his people. Connie told me that our world had been moved so that it could be hidden from the AOW. If they were to find out our world was not destroyed, as they think, we all could be in a much worse situation. Let us plan for this mission. It will take some time. That will be good for me. I need more time to heal."

Gunner, who had been listening to all of this, spoke up, "If all of this does not work, we need to be ready to take our people by force. I think we may need help from Terra's people, the Horsemen. They are excellent fighters. Our stun guns would work, but they are rather unimpressive. We need to create fear in the minds of those who resist. This should result in fewer casualties."

CHAPTER 60

Big Bad Surprise

"*ARMSTONG*, THIS IS the *Sally Ride.* We have landed. Proceeding to deploy forces to the south and west sides of the palace. Awaiting your confirmation that GK is in position to make an announcement over the loudspeaker, over," Gunner said.

"Copy that, *Sally Ride.* We have just now landed and are deploying forces to the north and east sides of the palace," Hatcher said.

As the men from both shuttles, over five hundred in total, began taking up positions around the palace, a third shuttle lowered a team of twelve special operations forces onto the large balcony on the third level, the level where the captive Earthers were located. The plan required that the twelve special ops forces break through into the rooms where the Earthers were being held. They would then hold in place as negotiations would begin to arrange a stand down of all "rebel forces," those under the command of Ishmael. A breakthrough of forces into the palace was viewed as risky. Too many innocent lives would be lost, and it was scrapped on that basis. A breakout by the forces who recovered the captive Earthers was also viewed as too risky. Some of the rescued personnel would be endangered, and it might not succeed anyway.

In the end, the goal was to reinstall GK to his position as emperor. He had shown a willingness to work with the Earth people and Ishmael had not. One of his points of contention was that GK was "too friendly" with the Earthers. GK found this to be odd since nothing but good had come from their relationship with the Earthers. His life had likely been saved by them. They had swiftly moved him out of the hands of the angry horsemen who wanted to kill him and had brought him back to his kingdom. Adding to that, they were helping to "electrify" his capital city, something that he had never dreamed possible.

GK thought, *Why is Ishmael so against the Earth people? Terra had told me about the enemies of The Mothers, the AOW. Were there any AOW people left after The Mothers discovered them and forced them out thousands of years ago? Had any AOW people helped build the terrorist bomb that almost killed him? Was Ishmael working with them? It seems likely. Well, I am going to give him a chance to live. I will exile him and whoever is working with him to the southern continent. That should take care of the problem. I do not want to execute him. He is family, after all.*

* * *

"I just got word from Captain Simmons. They've located our people on the third level. They are holding their position. They said they encountered no resistance. Odd. Oh well, at least that part went off without a hitch," Gunner relayed to Hatcher.

"We're in position here, Gunner. Awaiting GK's ultimatum, over," Hatcher said.

On the shuttle *Armstrong*, Connie and Terra helped GK come up to the bridge, where he could make his demands through the outside speakers they had set up around the palace.

"Are you ready, GK?" Connie asked.

"Yes, I think. Well, here goes." GK sat as he prepared to speak. "My fellow citizens, this is your leader, Genghis Khan. I was injured in the blast, but I am recovering. I have come here to tell you that I am ready to resume my duties. I understand that my cousin, Ishmael, has assumed my office. This is not in accordance with my wishes. We have a succession plan, but that was not necessary. I have radioed in several times to tell those who picked up that I was still alive, albeit badly injured. I made it clear that Minister Chen was my choice to conduct the affairs of the kingdom during my convalescence. He was doing so but was murdered by poison. Dr. Sampson told me he was killed by oleander poison. She collected a swab sample from Ishmael, taken from a cup he drank from on the night he illegally assumed my position. The sample showed he had handled oleanders sometime in the previous twelve hours. He was involved in the murder of Minister Chen. I am asking for him to be brought to the front gate of the palace and turned over to my forces located there. He will receive a fair trial. His punishment will fit his crime. It is my wish that if he surrenders voluntarily and is found guilty, his life be spared, but he will be exiled to the southern continent, along with any others who may have aided him in this dastardly action. He has five minutes to turn himself in. If he refuses, we will enter the palace forcibly and take him. We ask that all other palace personnel exit the palace at the same entrance so that we may conduct a search of the palace for any evidence concerning the murder of Minister Chen and for any evidence concerning the explosion. I am your rightful leader and am asking for your cooperation."

A crowd of people from the city had slowly been gathering as the events unfolded. They heard GK's words and were waiting to see what would happen next.

After four minutes, Ishmael and another man exited the main palace entrance. The unidentified man spoke into a megaphone that he had somehow acquired.

"My name is Raphael. I am a representative of the AOW or the Association of Worlds. We are a freedom-loving and peaceful people. We are only interested in your freedom and well-being. I have been on this world for thousands of years. I am immortal. My colleagues will be returning to this world to help you overthrow the yoke of oppression that The Mothers have planned for all of you. The so-called Mothers, you see, are in fact my enemy and yours as well. They do not have your best interests at heart. They will enslave you. If you join this fight on the side of my colleagues, you will be on the winning side. We have great plans for your world. We will give you technology that will make your lives easier. The Mothers have never given you technology to improve your lives. We will do so. I have recovered a communication device that allowed me to communicate with my colleagues. They did not know your world survived the movement of your world to its present location. Now they know that it survived and that the people here need their help. They are on their way as we speak to liberate you from these evil, oppressive Mothers and their puppet regimes led by your fraudulent leader, Genghis Khan. Join me in resisting! Fight for your freedom and the freedom of your children."

Inside the shuttle, Terra and Connie looked at each other with incredulity written all over their faces.

"Is this for real? Did he just call down the forces of the AOW on this world?" GK asked Terra and Connie.

"GK, I have been looking into his mind. Yes, he did exactly what he said. I think you should know that it will take them many years to come here, unless they have acquired The Mother's map of the universe, which, if they have it, could drastically reduce the time needed for them to travel here. We need to apprehend him and find out if that is the case," Terra said.

GK spoke again through the microphone, "Apprehend both men. We will question them and put them on trial!"

CHAPTER 61

Strategy Session

"RAPHAEL, I AM Terra, The Mother of this world, but you knew this already. You know I can read your thoughts. Even so, I want to know things that might take me quite a while to find inside your memories. You can make this process easier if you answer my questions. By doing so, you will be allowed to live. I know you are protected against death through natural causes, but not through accident or intentional action. You and your cohorts will stand trial. The system of justice here will require the death penalty. Their method of capital punishment is most brutal and painful, as you are aware. I will intercede on your behalf if you cooperate.

"I want you to tell me the names of all those who were involved in the bomb attack. I want you to tell me what you have been doing here on this world. Lastly, I want you to tell me what the AOW has planned for this world. Do they know where all the wrinkles are? How long until they arrive at this world?"

"I will tell you nothing! Your organization is an evil group of women! We will show no mercy to anyone like you!"

"Raphael, I know you do not actually believe what you are saying. I *know* this. Remember, I can read your mind and heart. You are repeating what you have been told, but you do not actually believe it. Your progression through your lives does not need to involve heavy karmic action *if* you acknowledge that what you did is wrong and ask for forgiveness. Part of receiving that forgiveness and avoiding very heavy karmic experiences in your next incarnation is to do your best to restore what you can from your wrongdoings. If you answer my questions willingly, you will live, and your karmic load will be lessened. If you do not cooperate, you will die, as the evidence against you is damning. What do you say?" Terra asked.

For the first time during the interview, Raphael was silent and thinking about what he would do and say next. He began to sob quietly, then it became louder.

"OK, I will cooperate! I have been here so long, alone, all alone. I had a wife and children on our home world. They are gone now—dead, I am sure. It has been thousands of years. All I want now is to have a family again, to be a father and raise children! Can I even have that if I cooperate? I don't want to live alone in a jungle on the southern continent. I would kill myself if that were my fate! Can I reclaim myself? Will you forgive me for what I have done? I killed hundreds of people. Is there forgiveness for that?"

Terra listened, and now she too began shedding tears. "Yes, Raphael, there is forgiveness for all of our actions. Many centuries ago, on my world, when Christianity was the dominant religion, there were a number of Christian cults that taught that certain actions were either unforgivable or only partially forgivable. These types of teachings were wrong and simply the ideas of those who founded those cults. I can see your heart. I know you are sincere. We will spare your life, and you can live among us at Benston. The people here are younger souls. They may not believe your sincerity, but among the Earth people, you can find what you desire. All things are possible. Earlier Christian teachings denied the reality of living many lives, but that was a specific choice on the part of those cultists. They had greater control over people if they could frighten people into believing they had but one life to perfect themselves. You may yet suffer some karma for your actions, but you can find what you desire." As she said this, she embraced Raphael, who continued to sob uncontrollably, only now it was tears of joy.

* * *

Terra had remained behind in GK's kingdom. GK had named her as the regent in his absence. His repaired organs were just about ready to be put into his body, which meant he had to return to Benston. The radio connection between Benston and GK's palace was still working, so Terra would be able to participate in a strategy meeting that was planned to help prepare for a possible AOW attack.

Terra had summoned Eve to discuss what to do about the possible AOW invasion since they now knew that Preemus still existed and that another world, Terra Nova, existed and was being fostered under a Mother. Hiding these worlds from the AOW had worked for thousands of years, but now their existence and location were known to the AOW again.

After Eve appeared, she began, "The AOW will undoubtedly try to destroy our worlds, Preemus and Terra Nova. The last time they used asteroids flung from very far away to destroy our world. At that time, we had no remaining starships or other detection methods to detect any threats like that. Now we do. I cannot read the minds of any of the AOW top personnel, as I do not know them, and they are too far away from me to read anything. In order to ascertain what their plans are, we would need to get someone closer and perhaps on their world to know more. This is a threat to your world and ours. We must take it seriously. Anyone who tries to infiltrate their world would be in grave danger. It would have to be someone who knows about their world and another person, or persons, who have the gifts of The Mothers. Raphael would be a logical choice, but he has not lived there for thousands of years. Another 'angle' will have to be discovered that could provide a plausible reason for people to travel to their world. Perhaps Levvi Markanen, being a well-regarded physicist, could travel there with someone who has the gifts. He could claim to have knowledge of how to travel to other universes and times and was willing to sell it to the AOW. I will go with him. None of the AOW people alive remember me or what I looked like."

Terra looked concerned and asked, "Mother, can they harm you? What if you are captured?"

"The Ones will be with me. No harm will come to me or those who go with me."

* * *

The strategy meeting started at 10:00 a.m. local time in Benston on a Monday, a week after Terra had assumed the role of regent of GK's kingdom during his convalescence.

President Stanton brought the meeting to order, then he turned the time over to Terra, who was listening in on the radio. She had asked if she could speak first because she had information from Eve to share.

"Thank you, President Stanton. I spoke with Eve. She cannot 'see' what AOW leaders are thinking. The distance is too great, and she does not know who the decision makers are. She feels we should send some people to Kepler 452b and get a 'read' on what their intentions are. Of course, the distance from here to Kepler 452b is much greater than anywhere else we have traveled to before. The distance is 1,350 light-years. The distance from Earth to Terra Nova was fifty light-years.

We made that journey in twelve years because we had The Mother's map of the universe showing shortcuts through the wrinkles in spacetime. I think we need to run a stellar flight profile for a trip to Kepler 452b and see how long this journey may take.

"Also, as concerns our 'cover story' for showing up there, Eve suggested that she and Levvi could travel there ostensibly to 'sell' the knowledge about how to travel to other spacetimes. It might be of interest to the AOW. They did show up on the Earth in Levvi's spacetime, by accident, as the *Calypso* also did on their way to rescue survivors on the Earth and Mars."

Stuart stood and asked to speak. "I am Stuart Carroll. I made the trip back to Earth and Mars to rescue survivors. We developed a rudimentary cloaking system based on knowledge given to us by Captain Tennyson of the *Nautilus*. It worked and allowed us to destroy the AOW vessel that was following us. I have knowledge of these wrinkles and the issues involved in traversing them. I want to go with Eve and Levvi."

President Stanton turned to Captain Simmons, who was seated next to him. "Captain Simmons, we need an experienced captain to pilot the *Calypso* to Kepler 452b. Would you be willing to go?"

Captain Simmons did not reply immediately. Instead, she sat and thought for a moment. "OK, I will go, but I want my husband and children to stay here. It will be a dangerous journey. I understand what is at stake. It is our lives and our new home here on Terra Nova. I will submit a list of names of personnel whom I would like to come along and run the ship. They should all be volunteers. This is a risky journey."

* * *

Stephanie and Anna, with Stuart and Levvi's help, created a flight plan. Between Terra Nova and Kepler 452b, there were two giant wrinkles. Stuart speculated that the tremendous "open space" between two locations might lead to bigger wrinkles as the density of mass in those regions was lower, meaning that the wrinkles could have fewer restrictions to their growth as cosmic inflation occurred. Whatever the reason, these large wrinkles were potentially blessings and curses. They were blessings in that the bigger the wrinkle, the better the shortcut. The downside was that this meant the AOW, if they knew about the wrinkles, could also travel the reverse distance from Kepler 452b to Terra Nova quickly as well. Lastly, there was always more danger in traversing a large wrinkle. The entry angle had to be precise, or they could go back in time or even enter another universe. Stuart knew this all firsthand. At any rate, the voyage was going to take sixteen years in each direction of travel, or thirty-two years round trip. Those who went on the voyage would have to endure a longer time away from their spouses and children.

Terra ended up staying in GK's kingdom for almost four months. At that time, a shuttle returned to deliver

Connie and GK and to pick up Terra and return her to Benston. During the four months that Terra was in GK's kingdom, GK had been given his new organs.

When Terra returned to Benston, she eventually gave birth to another set of twins, a girl and a boy. Terra intuitively felt that her new daughter would have "the gifts." She also felt her son might as well. This was a rare occurrence for a male child to have the gifts, but it did happen. Her new children were named Melissa and Hatcher Jr., after his father.

Stuart and Levvi had received the stem cell serum that Timeye had developed, which he believed would lengthen their lifespan to the length of the Caldonians. Most of the colonists had been given the serum as well. The average Caldonian lifespan was just under seven hundred years. This was of great comfort for those who volunteered for the voyage to Kepler 452b, as they would still be relatively young when they returned to their loved ones. Terra had forbidden any Terran natives from receiving the life-lengthening treatments. Eve had told her they needed to live shorter lives so they could progress more quickly in their incarnations. The Earth people had already progressed through many lives and had experienced and learned much from their earlier lives.

CHAPTER 62

Farewell

A YEAR LATER, THE *Calypso* had been fully prepared for the long voyage to Kepler 452b. In that time, Timeye had improved the cloaking device and had built a laser defense system for the ship. They would not be going into the lion's den defenseless.

Two shuttles were appropriated to the *Calypso* so that they could get personnel to the surface of Kepler 452b. The shuttles were outfitted with cloaking technology so that they could land without being detected.

Timeye was greatly worried about Captain Simmons. They had four children who ranged in age from fourteen to three years in age.

"Timeye, my love, I know you will worry about me, but Eve has promised that we will all live to see our loved ones again here on Terra Nova. I do this so that our home here will be protected for you and our children. Take care of them. Let them speak to me often via EPC. It will be better than nothing. Do not overly worry about me. I'll be doing enough of that for both of us!" Captain Simmons said through tears.

"I will, my dear. I will think about you every day that you are gone. I love you with all my being!" Timeye said.

Eve had traveled to Terra Nova to join the rest of the crew. She and Stuart went everywhere together. They were inseparable. They both knew why this was. Stuart had been Eve's first husband thousands of years earlier. They were making up for all the time apart by being together all the time now.

Hatcher and Gunner were piloting the shuttle that would take the crew of forty-seven up to the *Calypso*. Terra was there to wish them farewell. For most of the colonists, this was the first time they had seen Eve of Eden, The Mother of their home world, the Earth.

"My dearest child, Terra. While I am gone, you will not be on your own. My children on Preemus will be there to help you. Sing The Mother's song as you would for me, and one of them will come to your aid. Farewell for now. We will meet again when we return," Eve said with strength in her voice born from countless trials and thousands of years of experience.

PART SIX

In the Lion's Den

CHAPTER 63

Riding the Big Wrinkles

A YEAR LATER, THE *Calypso* came to the first of the two large wrinkles. "Captain, we are ready to initiate the entrance protocol," Anna Stetson announced to Captain Simmons.

"Proceed, Ms. Stetson," the captain replied.

Eve and Stuart stood behind the captain, who was standing behind the seated Anna Stetson. The ship rumbled and vibrated like a car running over rumble strips. The transit time was an astonishing four hours!

Will the ship survive four hours of this vibration? The whole duration of our flight from the Earth to Terra Nova never had this kind of "turbulence." What kinds of things will come loose after four hours of this? Stuart thought.

"Captain, our acceleration is now at 20 percent of light speed. We're going to get a nice boost from time dilation at this velocity for four hours," Ms. Stetson observed.

"Yes, we will, Ms. Stetson. Yes, we will," Captain Simmons responded.

After four hours, the *Calypso* punched through to the other side of the wrinkle.

"OK, people, let's try to find out where and *when* we are. Stuart and Anna, please let me know when this can be verified. I'm going to get some rest. Stuart, you have the helm," Captain Simmons announced.

After the captain left, Eve said, "There is a world nearby that is inhabited. It has not been located by the AOW. There will be one other world that is also inhabited as we proceed to the second large wrinkle. If we needed to, we could stop there for anything we need. The peoples of these worlds are very advanced and very kind."

"Good to know! I think I need to go potty!" Anna Stetson announced with a grin on her face.

Eve smiled back at Anna, then said, "These worlds I spoke of are good examples of how a world is supposed to develop. They eliminated patriarchy very early in their development and have had only a few wars. They are now completely peaceful. Their entire worlds will become worlds of The Mothers very soon. We have not yet asked them to contribute to our work, but they are ready."

Levvi, who had entered the G&C room, said, "I think my Earth has actually had contact with one of these worlds in my universe. Of course, the messages took thousands of years to reach us. We could not decipher their language, but we knew we were talking to an intelligent race. I think they wanted to talk to us. I'm told most advanced worlds use EPC to avoid detection by other worlds just in case there are hostile aliens."

"Stuart, this is Buck. I've got something on the differential gravitometer. It's a ship. It is cloaked, but it is giving itself away because of its mass."

"Copy that. Captain Simmons, this is Stuart. We have detected a large vessel at about two hundred thousand kilometers. We are cloaked. Hopefully, they haven't detected us. What are your orders?" Stuart asked.

"I'll come back to manage this. Simmons out."

"Is there any way we can tell if this is an AOW vessel?" Captain Simmons asked.

"Well, it is much bigger and more massive than the ship we encountered on our way back to Terra Nova," Stuart said.

"Eve, can you tell us whether this ship is an AOW vessel?" the captain asked.

"It is not an AOW ship. It is from the first world I told you about. They have very advanced space vehicles."

Just then, the com engineer, Peter Lefoot, called in over the intercom. "Captain, I have a message from the vessel we have detected. It is in English. They are requesting permission to send a shuttle to our ship and discuss a matter of great urgency. I can broadcast your response. Let me know, over."

"This is Captain Sandra Simmons of the FBS *Calypso*. Identify yourself and state your intentions, over," Captain Simmons said.

"I am Captain Jah Richtop of the *Deranger*. We have just evaded an armada of AOW vessels. We request your help in returning to our world. Our gravimetric disguise devices are not working, over."

"Permission granted. Our shuttle bay is on our port side. Continue coms on this frequency," Captain Simmons said.

Twenty hours later, a shuttle craft entered the shuttle bay, and a group of people, who again looked like Earth people, came out from the strange craft.

"I am Jah Richtop, commanding officer of the *Deranger.* We need to talk about what we encountered. If your intention is to continue toward their home port, what you call Kepler 452b, you will want to hear what we know."

CHAPTER 64

Getting by with a Little Help from Friends

"CAPTAIN SIMMONS, THE AOW is putting implants into every person who lives on their world or visits it. Are you familiar with what I am referring to?" Captain Richtop asked.

"We are. We had two individuals who became prisoners of the AOW. They both had these ghastly devices implanted into their heads near their temple," the captain answered.

"I don't know if your intentions were to pay a visit to their world, but you should know you will never have your own thoughts and dreams again with this device implanted into your head."

"I see. What else can you tell me about their world, its defenses, and such?" Captain Simmons asked.

"I was surprised to find out that their ability to deter incursions into their solar system is not that great. We were able to get within one AU of their world using our cloaking device. We sent a shuttle farther. The shuttle was stealthy. It had a cloaking system. Given its small mass, it was not detected by their differential gravitometers. There are holes in their defenses. Our mission was to try to penetrate their defenses and put people on their planet to spy on them. They have not yet tried to take our world, but we believe they will eventually try. Our defenses are formidable, but also not perfect. We are in a standoff with them at this time. We have heard about your colony on the world you call Terra Nova. Your world has a Mother on it, we are told, hence the interest by the AOW in acquiring your new world."

"How did you hear about us?" Captain Simmons asked.

"We captured an AOW spy who dwelt on your original world, Earth. He was left behind when they all fled the destruction they unleashed upon your world. He found an abandoned AOW spacecraft and left the Earth, hoping to be picked up by his people. He had put himself into hibernation. When we found him, he had been in hibernation for more than ten Earth years. We knew he would have important information, so we questioned him. We have techniques for getting information that are not barbaric like their brain implants, but they work very well," Captain Richtop answered.

"Well, Captain, we will escort you back to your world. If we encounter an AOW vessel, we believe we have even odds at winning a battle," Captain Simmons said.

* * *

The discussion with the captain of the alien vessel yielded important results. They had replicated the brain implant device. The people they put on the surface of Kepler 452b had these imitation brain implants. They gave off the same electronic signature as the real devices and were good enough to fool the AOW scanners when they checked their people at random checkpoints. Captain Richtop gave Captain Simmons enough of their devices so that everyone on board the *Calypso* could have one implanted to help protect them from having the real devices implanted if they were captured by the AOW in space or on the surface of Kepler 452b. Captain Richtop also informed Captain Simmons that they referred to their world as "Freedom and Peace." The irony was not lost on Captain Simmons and other *Calypso* crew members.

"On our world, many autocratic regimes from our past would use names for their countries that were gigantic contradictions from reality. It appears that the AOW also employs such propagandist techniques!" Stuart noted with a chuckle and a smile.

"Indeed. As evil as they are, they are certainly capable and a foe worthy of respect. I support your efforts against them. I will not ask you exactly what you are doing for two reasons. First, I can always truthfully claim ignorance if these narcissistic bastards capture me. Secondly, I realize that operational security dictates that as few people as possible know about delicate plans. On another subject, your vessel is known to the AOW. There were survivors of the encounter you had with them on your way back to Terra Nova. The 'mugshot' of your vessel is known and is circulating. The AOW encounters our ships often enough and knows we have slightly superior technology. They leave us alone. My fear is they will nonetheless attack my world someday. Your enemy is my enemy, and we are therefore allies. We know how to evade their defenses. We can get you close enough for one of your stealthy shuttles to land unnoticed. Your shuttles are stealthy, correct?" Captain Richtop asked.

"Yes. We have equipped two of them with cloaking capabilities and other defenses that we choose not to discuss. They can defend themselves. Our mission is to gain access to AOW leadership and learn their intentions. Their intentions will be of interest to you. You stand in the way of them attacking the world of The Mothers and the new world, Terra Nova, which is being fostered by a mother. We have the capability to know their intentions if we learn who is actually in charge and are close enough to read his mind," Stuart explained.

"I will not ask if you have a Mother, but it certainly sounds as if you do. I now can believe more comfortably that you will succeed. Our spies have not been able to learn much. I will discuss this with our defense minister. I think we should give you all the support you need. You should remain on your vessel in orbit around our world. There are AOW spies on our world. Your presence would be noted," Captain Richtop said.

CHAPTER 65

Sneaking in the Back Door

"CAPTAIN RICHTOP, I thank you for your help. We will execute our entry plan now in our shuttle. We will send you updates via EPC as we get information, per our agreement," Captain Simmons said.

The shuttle had just four people on board. The other crew members stayed with the *Calypso*. Levvi, Eve, Stuart, and Captain Simmons had the fake implants in their heads. It gave off electronic signatures that mimicked the real devices but did not allow for pain centers to be stimulated.

The shuttle *Armstrong* was operating in stealth mode.

I hope the cloaking and anti-gravimetric defenses worked as advertised. I guess I can be confident in the technology that Captain Richtop and his people gave us. They had landed spies on the AOW world successfully. Although they hadn't received any useful information from their people. Was that because they hadn't been able to get any good intelligence, or was it because they were captured or dead? I guess I'm borrowing trouble.

We have Eve, and she is not concerned. But she is immortal, so why would she worry? She did say none of us would be harmed. She must have a reason for her lack of concern and her confidence, Captain Simmons thought to herself.

"Entering orbit now, captain," Levvi announced.

"Copy that. We have the coordinates of the capital city. How many orbits until we begin a descent to the location outside the capital city?" she asked.

"Just two," Levvi answered. "We will do a deorbit burn for ten seconds, then we will free-fall until we deploy a descent chute, but that will be brief—only thirty seconds—then we engage the spinners. This is the best we could do for a stealth landing. Let's hope it is good enough not to get noticed."

The strange world had almost the same gravity as Earth, but it was 60 percent larger in diameter than Earth. The density of the planet was lower than the Earth, making the mass almost the same as the Earth's mass.

"It is noticeably bigger than Earth. Weird that the gravity is the same because of the density difference compared to the Earth," Stuart said, almost as if he were thinking out loud.

"You're sure no one will recognize you?" Captain Simmons asked Eve.

"Only Kannon and Ish were immortal, and neither of them is here now. They are elsewhere sowing discord. Any others have passed on, and they would not retain memories of me anyway. The Ones have told me this."

The descent to the planet's surface was surprisingly uneventful. There was no detection of radar or other methods of detection seeking the shuttle. It was nighttime, so they were using infrared to determine a landing location. They chose one behind some low hills. There did not appear to be any inhabitants nearby. Captain Richtop had told them that no one lived outside of the cities. All activities, like farming, were performed by robots with almost no human oversight.

"Engaging planetary sampling system. Sit tight for a few minutes," Stuart said.

"OK, background radiation is slightly elevated, for some reason. We will be fine, but we wouldn't want to live here for years. The atmosphere is slightly lower in oxygen than on Earth or Terra Nova, but acceptable. The temperature is 21 degrees Celsius. We can exit when we are ready. I assume we want to look around while it is dark?" Stuart asked.

"Yes, let's see what we can see under the cover of darkness," Captain Simmons said.

Everyone except Levvi left the shuttle to look around. Using night-vision goggles, they saw what looked like agricultural fields some distance away. They did not see any roads or other infrastructure.

"Well, this is not what I expected," Stuart said.

"Why don't we see anybody or anything manmade except some agricultural fields?" Captain Simmons asked.

"Their world is very controlled. Everyone has a job that they *must* do. There is no dissent—none. No questioning, nothing. This is their plan for everywhere they take over. This is so similar to a regime on Earth in the twentieth and twenty-first centuries. That regime was North Korea. It was destroyed in the wars of aggression. The traitor and top leaders in his party, which were secretly allied with Russia, tried to do this in the United States as well, but their plans were thwarted by the second civil war. But we are mistaken if we think we were unnoticed. They know we are here," Eve said.

Just then, holes in the ground about one meter in diameter opened up, and robots came out and began moving toward them. The robots were made to look like humans. Stuart and Captain Simmons drew their stun pistols and were just about to shoot them when Eve spoke.

"Do not fire on them! Remain perfectly still. I will pacify them."

They lowered their weapons, and Eve moved toward one of them, holding her hands out, with her palms facing upward. The robots stopped moving and moved back to the holes they came from. They descended back into the holes.

Stuart and Captain Simmons exhaled deeply and audibly.

Eve said, "They believe we are not a threat. We will not be bothered by them anymore."

“We need to get moving and go into the city to the safehouse where Captain Richtop’s colleagues are. Let’s try to do it before sunrise,” Captain Simmons said.

“No, we will enter the city in daylight. If we try to sneak in at night, we will be apprehended. No one is allowed outside at night. We will wait here until full daylight, then we will go in, through the front door, so to speak,” Eve explained.

CHAPTER 66

Inside the Regime

IN THE MORNING, they joined a long line of people who seemed to appear out of nowhere who entered the city called "Freedom's Capital" through what appeared to be the main gate. The city had cement walls, approximately five meters high, around the entirety of the city. The entering people seemed to have jobs in the city and came from the surrounding area. No hover cars were seen. Everyone walked or sat in the back of wagons pulled by, of all things, horses!

The Earthers had tried to dress casually so as to be unnoticed. Even so, they found their clothes stood out as a big step up in décor and quality from those they entered with. An automated robot sentry asked them their business in the city. They answered that they were visiting relatives and had the profession of teachers. No one had papers of any kind, but each entering person had to look into a device that presumably identified them by their iris. When it came their turn, the Earthers were flagged and told to proceed to an identification area.

"Who are you? Why are you here? Where are you from?" the armed interrogator asked each of the Earthers.

"I am Eve of the planet Albanus, your neighbors. I am here to visit friends, but in reality, we wish to sell important technology to your regime. Technology that will be of great interest to your world in its efforts to expand into new worlds, even worlds in other spacetimes. We ultimately wish to speak to your leader and his science advisor."

"And these other two? Are they with you? Are they here for the same purpose?" the interrogator asked.

"Yes. This is Captain Simmons, our ship captain, and this is Stuart Carroll, the chief science officer on our ship. He is a physicist by training. These two were on the ship that carried Earth people to the planet Terra Nova. They have experienced travel to other times in our universe, and they have traveled, albeit accidentally, to another spacetime and visited the Earth in that spacetime. We have another individual named Levvi Markkanen, also a physicist who developed the theories and equations for traversing wrinkles into past times and into different universes, or spacetimes. He is not with us. He stayed behind on our ship. He would like to enter the city eventually as well," Eve explained to the highly skeptical interrogator.

The interrogator looked at each of them for a moment, then huffed and headed off to the other side of the large interrogation room to speak to someone who appeared to be his supervisor. They spoke for several minutes, then he returned.

"I am to take all of you into custody. People from the Earth ship *Calypso* are wanted as criminals of the AOW

Empire. They attacked and destroyed an AOW ship! Guards, restrain these people!" the interrogator shouted.

"I hope you know what you are doing, Eve," Captain Simmons said quietly to her as they were handcuffed.

"I do. Do not worry," Eve answered softly.

* * *

As they walked through the city to the detention center, they suddenly stopped. The guards ordered them to bow their heads as a large procession proceeded down the street.

"All hail, Leonard the Fourth, Supreme Leader, Conqueror of the Galaxy, the Most Benevolent, Wisest, and President of the Quorum of the Twelve Apostles and God of our Galaxy!" a man at the front of the procession announced.

"All hail, Leonard the Fourth," shouted the people on the street as he passed by.

The guards poked the Earthers in the side with their stun sticks and told them to repeat what the people said.

"I take it humility is not his strong suit?" Stuart asked the guards who had smacked him in the mouth with their rifles.

"Relax, ladies! Would the most benevolent be pleased with your actions?" Stuart asked. This time, they hit him over the head, causing him to fall to his knees.

Stuart got up and glared at the guards. Eve whispered to him to be quiet and let her handle things.

They eventually reached a compound that was surrounded by another wall with razor wire on top of the wall. They were led inside and placed in separate cages that were adjacent to each other. The cages were in the outside courtyard of the prison, exposed to the elements.

"We really do have technology that your world would be interested in. Your leader may not appreciate your inattentiveness to this most generous offer we wish to make. We are offering to give the technology to your empire, for free. But if Stuart is killed or suffers irreversible brain injuries, he may not be able to share the technology, and that may look very bad for your advancement opportunities. Your choice," Eve explained, smiling. The guards hesitated, then they left and spoke to what seemed to be their supervisor.

After several hours, in the late afternoon, several men, dressed much better, approached the cages and removed all three prisoners and put them into a waiting hover car.

The car drove through the city and then left via the front gate. It continued driving for over an hour at high speed. Eventually, it slowed to a stop as they came to what had to be the grandest palace they had ever seen. The car was ushered through two separate gates in two different walls that surrounded the gaudy and extravagant palace. It stopped outside an entrance to the palace building itself. They were ordered out of the hover car and brought inside, where a man greeted them, introducing himself as Mr. Kim, the science advisor to the supreme leader. He led them into a room where they were seated. A few minutes later, the supreme leader entered.

CHAPTER 67

A Conversation with the Supreme Leader

THE SUPREME LEADER looked like a woman in drag, Stuart thought. *What's with all the eyeliner? And the long dark hair, carefully braided and wrapped on top of his head. The "imperial robes" were quite beautiful. His fingernails were ridiculously long. His skin was pale, soft, and supple. Yes, he looked almost feminine. Why would he go for this look on a world supposedly ruled exclusively by men? Were all the men here so "feminine"? He didn't say anything, so his voice was a mystery. Was it a high-pitched voice?*

An aide spoke. "The Supreme Leader welcomes you to his humble palace. His Holiness is most busy. This audience with His Holiness is a sign of his magnanimous nature. He wishes for you to speak and be brief. He wants to know of this technology you wish to share with his subjects. Speak now."

"Your Holiness," Eve stood as she began speaking. "Your Holiness, my colleagues and I wish to share hidden secrets of 'The Mother's map of the universe.' I understand you have been seeking this map, as the possession of it would open the possibility of traveling throughout our galaxy and, indeed, the universe. The premise of 'wrinkles" in spacetime is, I am sure, known to his holiness. "Traveling through the wrinkles shortens travel time between locations drastically. Indeed, it makes interstellar travel a practical possibility, not something that takes many lifetimes to accomplish. But there are secrets to these wrinkles, especially the large wrinkles that have only recently been discovered. These discoveries were made accidentally by an Earth ship and an unfortunate AOW ship that pursued the Earth ship *Calypso*. The *Calypso* discovered these secrets on their trip back to planets Earth and Mars to recover survivors there after The Mothers ruthlessly destroyed the Earth out of frustration, something your holiness would never do." Eve paused and watched as the Supreme Leader looked over at his chief aide with a questioning look.

"We shall discuss this later, your Holiness," the aide coughed nervously and audibly in response to the unspoken question. "Please continue, madame Eve of Albanus."

"Yes, yes, of course," Eve smiled wryly. "As I said, these discoveries were 'accidental.' The first accidental discovery occurred while traversing a large wrinkle and resulted in the *Calypso* traveling back in time some twelve or thirteen thousand years. In fact, the *Calypso* visited the ancient Earth and visited The Mother of that world, who gave the Earthers instructions on how to get back to their time frame. Unfortunately, they did not follow her simple instructions, and in attempting to return to their time frame, they entered a completely different spacetime, or universe. That universe had diverged from ours several hundred years earlier. Their bumbling was again mitigated, this time by assistance from a more advanced Earth. In fact, the brilliant physicist who discovered how to travel to other universes is our colleague, Dr. Levvi Markkanen. He is not with us today. He stayed to watch over our landing craft.

"He, or Stuart," Eve paused and pointed at Stuart, "can explain the detailed equations involved in precisely entering these large wrinkles so that the desired universe can be traveled to. There is, however, some caution that needs to be exercised in traveling to other universes. The Earth in the universe that the *Calypso* traveled to was sympathetic to the Earthers from our universe. They are much more advanced and were able to easily destroy the AOW ship that they followed back to our universe. They are a powerful ally of The Mothers at this time. I have heard it said they have a mutual defense arrangement with The Mothers, which states that an attack against either Earth or their allies requires the other side to render assistance against their common enemy, the AOW. So perhaps traveling to that universe is not in your interest, your Holiness."

No one knew it, but at that moment, Eve "pushed" the idea deep into the fear center of the Supreme Leader. She noticed that his facial expression showed concern at that moment. She also read his mind and determined that he, at that moment, decided against using this technology to expand his empire. Eve reinforced his fear and strengthened the idea that he needed to stick to smaller gains against The Mothers, lest they inform their more advanced and powerful allies to intervene on their behalf. The Supreme Leader did, in fact, know about the destruction of his ship by the Earth ship from the other universe. To that point, they had always had a technological advantage over any foe they encountered. That was no longer the case.

The Supreme Leader whispered something into his chief aide's ear. At that point, the aide said, "We will consider your offer. You reside on Albanus. We know where to contact you should we desire to do so. You are free to leave. We wish only peaceful relations with The Mothers. If you encounter them, please convey that to them."

And with that, the meeting was ended.

CHAPTER 68

The Journey Home

"WELL, THAT ENDED up going much easier than I thought it would," Stuart said to those assembled in the conference room on the *Calypso*.

"Yes, it really did work out well. The Supreme Leader had genuine fear in his eyes when he was told that the other Earth was now our ally and would come to our aid if we were to be attacked," the captain said.

There were nods of heads from those seated in the conference room.

Eve did not smile or join in the sharing of confidence. Stuart noticed this and asked, "What's wrong, my love?"

"We have only delayed an inevitable confrontation with the AOW over my world and Terra Nova. They will advance their technology and will become emboldened at a future time. Having placed fear in their minds, they will come to resent being made to look weak. I fear that we only delayed the inevitable. We shall someday fight these same people and their leaders again, only it will be a real confrontation next time, where lives, many lives will be lost," Eve said, looking very sullen and downcast.

“Well, at least we did buy some time. We can head home now. I want to get home and see my children and my husband,” Captain Simmons said, more soberly this time.

“Here, here! Here’s to returning home!” Levvi said. “I feel as if I have some unfinished business at home.”

PART SEVEN

Becoming the Goddess of Terra Nova

CHAPTER 69

Booms and Shaking

IT WAS A day no one would ever forget. Everyone would forever remember where they were and what they were doing when "it" happened.

The day started out like any other day. It was the transition between the rainy season and the dry season. That morning it rained, then it stopped, but it was quite cool. The ocean looked placid and peaceful. Around ten o'clock in the morning, a sound was heard. The colonists thought it sounded like a large rocket taking off in the distance, except that about a minute later, the ground moved so violently that nearly everyone was thrown off their feet.

In the governance building, which housed the congress when they were in session, a meeting was ongoing. Oddly, the discussion among the scientists was about putting in a series of seismographs so that they could begin to understand the seismology of Terra Nova. To that point, this had not been an issue that the colonists felt was important enough to deal with.

"Ed, that was a big earthquake. Should we be concerned about a tidal wave?" Dr. Ginger Kimball, a refugee from Mars, asked Edwin Godwin, a planetary scientist.

"We are in the dark on this. We don't know if this occurred somewhere on land, in the ocean just offshore of land, or if it was a volcanic eruption. But we need to get the shuttles up and have a look at the coastline across the Terran Gulf to see if there are tidal waves hitting those coastlines. We probably should put a shuttle up in orbit to see if there is anything to be seen indicating a volcanic eruption. We need eyes overhead to even begin assessing the threat," Edwin said, clearly shaken over the possibility of an even greater catastrophe looming over their heads, perhaps only minutes or hours away.

Just then, Hatcher, Levvi, Peter Chan, and Jacob came into the meeting room.

"I think we need to get the shuttles up to see if there is any sign of tsunami waves hitting coastlines far distant from us," Hatcher said to those in the room, not waiting for anyone to ask his opinion.

"We've just asked ourselves that question, Hatcher. We have six shuttles. I can make a quick and dirty guess as to where they should go to assess this situation. However, one or two of the shuttles need to go into orbit to look for any terrestrial volcanic activity. We will deploy four shuttles to coastline locations, and two in orbit. Here, let's do this quickly. Someone needs to get on the colony's giant voice and tell the colonists to gather supplies and be ready to evacuate to the coastal hills in case we detect a tsunami. Our elevated location might be enough to ride out a wave, but Ammonsberg is on a low-lying flat plain. They will need to move to higher ground. We could have their women and children come inside the Benston walls, but there probably isn't room for all the Terrans here in Benston. We need to tell the adult male Terrans to start moving up the hillsides. Get the hover cars and send someone over to tell them to gather a few supplies and get everyone moving pronto toward the hills or over here. We may have hours, or we may have much less time. Everyone needs to understand that there could be a series of waves coming. Do not leave the hills after an initial wave hits. More could come, even hours later," Edwin said.

Gunner, who had also entered the conference room, sat and listened to Edwin. When he was finished, he said, "I'll get a couple of security folks to go in the hover cars and warn the Terrans. I'll need to pilot a shuttle. I volunteer to go into orbit. Hatcher will take another shuttle and go into orbit, also. We will plan for our orbital placement to be 180 degrees opposite so we can get good and continuous coverage on the surface."

* * *

"Gunner, how do you read me?" Hatcher asked over the radio.

"I read you five by five. Hatcher, we will be executing an expedited dust-off. We've done this before. I know you have your NAV computer loaded with the orbital placement. I've got mine loaded up as well. You launch first. I'll wait thirty seconds, then launch next. You have Peter Chan to do your observations; I've got Dr. Ginger Kimball. She'll be my eyes looking down. Good luck! Godspeed."

"Three, two, one, ignition, liftoff," Hatcher said over the radio, which was transmitted back to the colony com room manned by Sam and another com engineer.

Flames erupted from the shuttle nozzle as the shuttle accelerated rapidly and was gone from sight in just one minute. Thirty seconds later, Gunner's shuttle lifted off.

Moments later, each of the remaining shuttles lifted off using their rockets to achieve a ballistic trajectory that put them very quickly in a location to observe four far-distant coastal locations. Their eyes in the heavens were on their way.

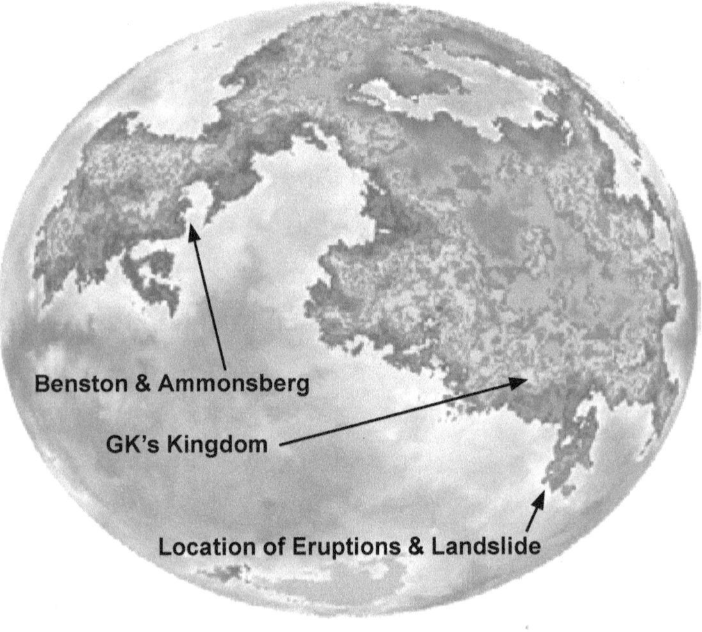

The Western Hemisphere of Terra Nova

CHAPTER 70

Eruptions and Waves

AS HATCHER'S SHUTTLE reached orbit, Peter began looking downward at the surface. The shuttle was orbiting opposite the direction of the planetary rotation, so each orbit would pass over the same land mass more quickly. On the first orbit, he saw a huge plume of smoke spreading from a location on the southernmost tip of the largest land mass. It had to be a volcanic eruption, Peter concluded. As he looked, a fireball erupted and came through the ash cloud. He had just witnessed a second great eruption of this volcano.

"Hatcher, I see a large smoke cloud on the southern tip of the kingdom of GK. Actually, it is an island at the southern tip. I'm getting a video, which I will send down to Benston. While I was watching, a fireball erupted through the ash cloud. I think this was another eruption. Whoa! Now I see a second and third location of fireballs near the first one. There might be more than one volcano that is erupting."

"Copy that. I'm looking at your video feed. The sound we heard must have been from the first eruption. Somebody on the ground can calculate the exact time of the eruption based on when we heard it and the distance between our location and the site of the eruption. The sound waves traveled at the speed of sound. That could allow us to estimate the time of arrival of a tsunami, which would travel very fast as well, if one is coming."

"What about waves hitting shorelines? Can you zoom in on the telescope and see any evidence of shoreline disturbances?" Hatcher asked.

"I'm looking at shorelines near the location. I might be seeing some evidence. The vegetation is wiped clean on shorelines nearby. We could compare the images I'm seeing now with the images when I did the first survey from the *Calypso*. I wonder if there was a landslide from the eruption, and maybe that caused the tsunami, or maybe an offshore fault ruptured coincident with the eruption and caused the tsunami. From what I remember, there was a lot more vegetation along this southern coast. We're passing out of view of our hemisphere now, so I won't be able to observe the coastlines farther out from the eruptions until our next orbit."

* * *

"Gunner, I assume you saw our video feed showing the location of the super-eruption. Did you observe anything in the other hemisphere?" Hatcher asked.

"Negative, all quiet. I think we have the smoking gun or mountain!" Gunner responded.

Just then, Peter started talking excitedly.

"I think I am seeing waves strike in real time south of our location at Benston. We're going to get hit real soon! I hope everyone reached higher ground. There isn't much time now. The waves are moving inland across the entirety of the large island south of our location! The elevation there is only thirty meters or so. I think our elevation at Benston is sixty or seventy meters. Maybe Benston will be spared. I don't think Ammonsberg is going to be spared, though!"

"Ground control, what is the status of evac out of Ammonsberg?" Hatcher asked.

"Not sure. Last I heard, a large group, mostly Terran males, had reached the foothills. I will get in communication with them. I'll check. Over," Benston coms responded.

"Peter, what do we do now? I don't feel good going back down and trying to land while they are dealing with all this. I hope we have a place to land!" Hatcher said.

"Well, the landing field is large and at roughly the same elevation as the rest of Benston. Hopefully, it will be there. In the worst case, we could orbit and rendezvous with the *Calypso* when it passes back our way. I don't think shuttles will be much use in search and rescue. They're too big and don't have easy egress for rescues from the ground."

CHAPTER 71

Becoming the Goddess of Terra Nova

THE EVACUATION OF Ammonsberg did not go well. The small dirt road between Ammonsberg and Benston was clogged with wagons and horses, and everyone was in a panic, hindering the movement of people. The male Terrans were hoofing it overland to the coastal range, several kilometers distant. That was also slower than anyone had imagined. The reports that the men had reached the foothills were simply not accurate. No one was going to reach safety. There had been no planning for this type of catastrophe.

Terra had sent little Hatcher and Melissa on a hover car to Benston, along with as many other children as they could get into the car, but there were still many children who were not going to make it out of the path of the waves.

Willie, who was at Benston when the ground started shaking, was now with Terra, trying to help people get out of Ammonsberg. Now, both of them were caught in no-man's land on the road between Ammonsberg and Benston. Terra was in a heightened state of fight or flight, but somehow she had managed to stay outwardly calm. Everyone was looking to her to do something, anything.

Just then, they looked out at the ocean and saw a stunning sight. The ocean had suddenly withdrawn from the shoreline. Fish were seen flopping on the muddy ocean floor. Pieces of timber from sunken ships stuck up from the mud. A rushing wind blew from the mountains toward the sea. An eerie silence followed, then everyone began pointing at the horizon. At first, people thought a cloud bank was moving toward the land from the sea. A few seconds passed, and a huge whooshing sound filled everyone's ears. They realized this was the first wave of the tsunami—a gigantic wave over thirty-five meters high was roaring toward the shore at an impossible speed. No one thought water could move that fast. People fell to the ground in fear. Crying and screaming could be heard everywhere. They were all going to die.

Terra saw all of this. Levvi, who had just returned from piloting one of the shuttles, saw it also from the Benston wall. He knew Terra was caught on the road with the Terran women and children. He left the main gate and went running toward the mass of humanity on the road.

Terra stepped from the roadway and walked quickly toward the shore. Some of the people shouted at her to come back. Terra ignored them. She had a singular purpose. She just wasn't exactly sure what that purpose was. She knew she needed to defy the wave somehow. She had to stop it. How she was going to do that, she did not know, nor did anyone else who saw this.

On top of one of the observation posts on the Benston wall, President Katherine Swinton observed all of this. She watched this through binoculars. She saw a woman calmly walking toward the sea as the sea sent certain death toward her and everyone else outside the walls of Benston.

Terra stopped where the water had been moments earlier, and she began to sing The Mother's Song. She finished and raised her arms up toward the sky and said, "I call upon the mighty winds to push this wave backward. I call upon the ground to heave upward and stop this wave!"

As she said this, a roaring wind knocked everyone standing to the ground. The wind continued to blow harder and harder, then the ground below their feet rumbled like a roaring freight train and lifted upward from the old shoreline to where the roaring wave of death was. The wave was slowed by the mighty wind as it crashed upon the wall of land, which had been uplifted in front of the wave. The wave beat upon the uplifted land and was stopped in its tracks. It spread outward all along the length of the uplifted land and recoiled back out to the open sea.

Everyone watching saw this, including Levvi, who had just arrived and stood behind Terra. The crying and screaming stopped. Silence filled the air, which, only moments before, was filled with screams and the sounds of a wave of death roaring toward them. People began walking toward Terra, wanting to touch her. Levvi put his arms around Terra and hugged her tightly. They had witnessed an otherworldly experience and were in a state of shock. One moment, they were certain they were going to die, and a moment later, they realized they were going to be spared. No one had the slightest explanation for what they had seen with their own eyes.

They had known that Terra was special, that she was kind, and cared for them. Now they looked at her in a completely different way. Many of them dropped to their knees in front of her and began to shed tears as they thanked her for saving their lives.

They now viewed her as a *goddess, the goddess of Terra Nova, The Great Mother of all who lived or who would ever live on their world.*

The End of Book Two

www.ingramcontent.com/pod-product-compliance
Lightning Source LLC
LaVergne TN
LVHW100136061126
851857LV00010B/105
* 9 7 9 8 9 0 1 2 4 2 8 6 5 *